Elliptical Fiber Waveguides

For a complete listing of the *Artech House Optoelectronics Library*,
turn to the back of this book.

Elliptical Fiber Waveguides

R. B. Dyott

Artech House
Boston • London

Library of Congress Cataloging-in-Publication Data
Dyott, R. B. (Richard B.)
Elliptical fiber waveguides / R. B. Dyott.
Includes bibliographical references and index.
ISBN 0-89006-477-6 (acid-free paper)
1. Optical fiber detectors. 2. Optical wave guides. 3. Dielectrical wave guides. I. Title.
TA1815.D96   1995   94-44501
621.36'92–dc20   CIP

British Library Cataloguing in Publication Data
Dyott, R. B.
Elliptical Fiber Waveguides
I. Title
621.381331

ISBN 0-89006-477-6

© 1995 ARTECH HOUSE, INC.
685 Canton Street
Norwood, MA 02062

International Standard Book Number: 0-89006-477-6
Library of Congress Catalog Card Number: 94-44501

10 9 8 7 6 5 4 3 2 1

To Janet

Contents

Preface

The concept of communication using optical fibers was the long-sought-after solution to the eternal problem of how to carry ever-increasing amounts of information over ever greater distances. The fiber that eventually evolved fulfilled these two main needs: low loss and high bandwidth. By contrast, the much broader field of optical fiber sensors needs fibers with a more diverse range of characteristics.

Evolution has been towards encapsulating the light paths in optical interferometers (Michelson, Mach-Zehnder, Sagnac), to act as compact and stable elements that use phase rather than amplitude as the sensing parameter. Each path in such a sensor has to be a single path, and so the need for a truly single path fiber that holds the polarization of the light it guides.

The majority of polarization-holding fibers have worked on the principle of creating two decoupled paths, changing the index of the fiber material anisotropically by using stress; the stress being created by the difference in thermal expansion of elements built into the fiber. There are, however, some advantages to using the geometry of the waveguide to split the fundamental mode into two orthogonally polarized modes. Such geometrical birefringence is a characteristic of a fiber that has an elliptical section.

One of the advantages of the elliptical fiber is the simplicity of its construction and, therefore, its ease of manufacture compared with that of the stress-induced birefringent fiber. A further advantage of not using stress is that sensitivity to temperature and strain is greatly reduced, an important consideration with interferometric sensors. Another interesting property is that the high-order modes are azimuthally stable compared with those in a circular core fiber, and this leads to the possibility of over-moded fiber sensors—a major application.

Thus, there is an increasing interest in elliptical fibers not only for sensors, but for other more recent applications such as rare-earth-doped fiber sources and amplifiers or

---

even in communications for neutralizing group delay. This book is written as an introduction to the subject.

I would like to thank my colleagues at Andrew Corporation for their help in gathering data and Professor Brian Culshaw for encouragement and criticism. I am particularly grateful to Tom Monte for very many helpful discussions and for advice on the text. The support of Andrew Corporation is gratefully acknowledged.

Chapter 1

Introduction

1.1 HISTORICAL INTRODUCTION

The history of dielectric waveguides can be traced back to 1899 when Sommerfeld [1] published an analysis of the propagation of electromagnetic waves along a conducting wire surrounded by a dielectric. He showed that a surface wave of low attenuation could be propagated and that the radial extent of the field and the wave velocity depended on the conductivity and radius of the conductor. In 1907, Harms [2] extended the investigation to a wire of infinite conductivity coated with a layer of dielectric of higher permittivity than the surrounding medium. Many years later, in 1950, Goubau [3] gave details of experiments with such a waveguide, the surface wave transmission line, and showed that the transmission of power over comparatively long distances over a wide bandwidth was possible.

The guiding of waves by a dielectric rod of circular cross section was first analyzed by Hondros and Debye in 1910 [4] and 1911 [5] with an experimental confirmation by Zahn [6] in 1915. However, unlike metal-walled guides, both the dielectric rod waveguide and the surface wave transmission line have a radial evanescent field that causes problems with such practical matters as supporting the guiding structure and protecting it from the environment. Moreover, the dielectric guide is too lossy to be used as a long-distance transmission line although such a system, using ultra-low-loss dielectric was tried out by the British Post Office in the 1960s. Thus, interest in the surface wave transmission line faded and the main use found for the dielectric rod was as an antenna [7].

In 1961, Snitzer and Osterberg [8,9] applied the theory of the dielectric waveguide to optical fibers where the evanescent field is enclosed in a glass dielectric cladding, thus solving the problem of protection and support, and in 1966 Kao and Hockham [10] and Wertz [11] almost simultaneously proposed the optical fiber as a long-distance transmission line for communications. At the time, the other problem still remained since fibers made out of the most transparent optical glass had an attenuation of about 600 dB/km, later reduced to 6 dB/km by meticulous work at the British Post Office. It was not until the key achievement by Kapron, Keck, and Maurer at Corning [12] in 1970 of a process for making low-loss silica fiber that the dielectric guide came into its own. With the small difference in refractive index between the central core and the surrounding cladding, which is characteristic of an optical fiber used for communications, it became possible to simplify the analysis. A major contribution was the reclassification of waveguide modes into linearly polarized groups by Gloge [13].

The first reference to elliptical waveguides seems to have been in 1938 when L. J. Chu wrote his doctoral thesis on electromagnetic waves in elliptical metal pipes, which appeared in the *Journal of Applied Physics* [14] in the same year. In 1961, the first solution for a dielectric rod of elliptical section was published by Lyubimov, Vaselov, and Bei [15]. Other extensive analyses for the elliptical dielectric waveguide by Yeh [16] and Piefke [17] followed in 1962 and 1964.

One of the early concerns with circularly cored fiber for communications was the effect of deviations from circularity on the system bandwidth caused by the two orthogonal fundamental modes of a slightly elliptical guide traveling at different group velocities. Thus, the elliptical fiber was first analyzed as a nuisance by Dyott and Stern [18], and by Schlosser [19]. Later, it was realized that if both the ellipticity and the index difference were made sufficiently large, the propagation constants of the two fundamental modes could be separated so that the intermode coupling would be small enough for a fiber to hold polarization over a considerable distance (Ramaswamy et al. [20], Dyott et al. [21]), a useful attribute for fiber interferometers. Recent applications to fiber sensors have used the elliptical fiber in the higher order mode region where the modes are azimuthally stable compared with those in a circular fiber (Kim et al. [22]).

1.2 NOMENCLATURE OF MODES

The usual system for labeling modes in metallic waveguides is to designate the mode by the field which is purely transverse to the direction of propagation, that is, transverse electric (TE) or transverse magnetic (TM). An alternative system, which has fallen out of use, is to designate the mode by the field which has a component in the direction of propagation, that is, electric (E) or magnetic (H). For metallic waveguides, with the exception of the surface wave transmission line, either system will do, but for the hybrid modes on dielectric guide neither field is purely transverse and there is a component of both fields in the direction of propagation. It is therefore necessary to use the second system, calling the modes EH or HE and putting the dominant field first. For the circularly symmetric modes, which have a purely transverse field, it has been the custom to slip into the other nomenclature and use TM or TE rather than E or H. Thus, the hybrid modes on the circular dielectric guide are labeled HE_{nm} and EH_{nm} and the circularly symmetric modes TM_{nm} and TE_{nm}. There seems to be no reason to use this mixture of nomenclatures which can be confusing with its dual concepts of fields transverse to, or in the direction of, propagation, and so the E, H system will be used throughout the book.

1.3 DIELECTRIC WAVEGUIDES

The basic dielectric waveguide consists of two regions: (1) an inner region, or core, of dielectric constant ε_1 (refractive index n_1) surrounded by (2) an infinite outer region, or cladding, of lower dielectric constant ε_2, (n_2) (see Figure 1.1).

The discontinuity at the core-cladding boundary means that functions describing the fields inside the core must be quasiperiodic. The outer boundary for the cladding is at infinity where the fields must die away to zero, so that the corresponding functions are quasiexponential. The periodic nature of the fields in the core allows the existence of different modes, each with its own field pattern; the number of allowable modes increasing with the dimensions of the core.

The argument of the functions describing the fields in the core is designated as u, and in the cladding as w, with u and w related to the propagation constant $\beta = \dfrac{2\pi}{\lambda_g}$ by

$$u^2 = k_1^2 s^2 - \beta^2 s^2 \tag{1.1}$$

$$w^2 = \beta^2 s^2 - k_2^2 s^2 \tag{1.2}$$

$$V^2 = u^2 + w^2 = k_1^2 s^2 - k_2^2 s^2 \tag{1.3}$$

with

$$k_1 = \frac{2\pi\sqrt{\varepsilon_1}}{\lambda_0} = \frac{2\pi n_1}{\lambda_0}$$

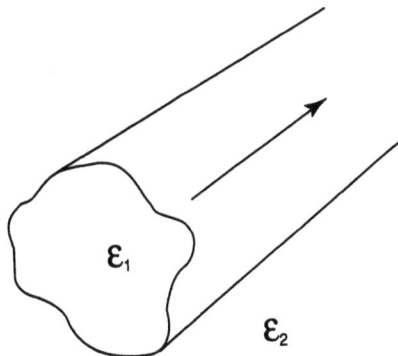

Figure 1.1 Dielectric waveguide—general concept.

and

$$k_2 = \frac{2\pi\sqrt{\varepsilon_1}}{\lambda_0} = \frac{2\pi n_2}{\lambda_0}$$

as the wavenumbers in the core and cladding. The variables λ_g and λ_0 are the guide and free-space wavelengths, and s is the semilateral distance of the core (e.g., the half-width of a slab, the radius of a circular rod, or the semiminor axis of an elliptical rod).

The variable V is often called the normalized frequency and is an intrinsic parameter of the guide.

$$V = ks(\varepsilon_1 - \varepsilon_2)^{1/2} = ks(n_1^2 - n_2^2)^{1/2} \tag{1.4}$$

where

$$k = \frac{2\pi}{\lambda_0}$$

is the free-space wavenumber.

For hollow metal waveguide at microwave frequencies $\varepsilon_1 = 1$, $\varepsilon_2 = 0$ and V becomes

$$V = ks$$

The geometry of the waveguide section governs the coordinate system into which Maxwell's equations are cast. The resultant differential equations determine the types of functions F_{periodic} and $F_{\text{evanescent}}$. The solutions of the waveguide characteristics are then obtained by matching the tangential fields at the core-cladding boundary to produce the transcendental equation

$$\left(\mu_1 \frac{F'_{\text{per}}(u)}{uF_{\text{per}}(u)} + \mu_2 \frac{F'_{\text{evan}}(w)}{wF_{\text{evan}}(w)} \right) \left(\varepsilon_1 \frac{F'_{\text{per}}(u)}{uF_{\text{per}}(u)} + \varepsilon_2 \frac{F'_{\text{evan}}(w)}{wF_{\text{evan}}(w)} \right)$$

$$\tag{1.5}$$

$$= n^2 \left(\frac{\mu_1}{u^2} + \frac{\mu_2}{w^2} \right) \left(\frac{\varepsilon_1}{u^2} + \frac{\varepsilon_2}{w^2} \right)$$

with the prime denoting differentiation with respect to the argument. This equation applies to all isotropic waveguide compositions and for any geometry where the variables are separable.

The two bracketed terms on the left-hand side represent solutions of the modes sensitive to the permeabilities μ_1 and μ_2 and sensitive to the dielectric constants ε_1 and ε_2 of the core and cladding, the μ-sensitive modes having re-entrant electric fields with no charge on the boundary (e.g., the H_{0m} modes in circular core guide).

The mode types are coupled by the right-hand side to form hybrid modes HE and EH, which have both E and H fields in the direction of propagation. The variable n is the number of field variations in the azimuthal sense in circular and elliptical guides and in the transverse sense in rectangular guides. When n is zero, as with the infinite slab or the circular symmetric E_{0m} and H_{0m} modes, the modes are decoupled and have only either E or H fields in the direction of propagation.

A dielectric waveguide with an elliptical cross section turns into an infinite slab at one extreme and a circular rod at the other. The slab and the rod are both simplifications of the more complicated elliptical section and so they will be treated first.

REFERENCES

[1] Sommerfeld, A. "Uber die Fortplanzung Elektrodynamischer Wellen langes eines Drahtes," *Annalen der Physik und Chemie* (New Series), Vol. 67, 1899, p. 1.

[2] Harms, F. "Electromagnetic waves on a wire with a cylindrical insulating sheath," *Annalen der Physik,* Vol. 23, 1907, p. 44.

[3] Goubau, G. "Surface waves and their application to transmission lines." *Journal Applied Physics,* Vol. 21, 1950, pp. 1119–1128.

[4] Hondros, D. "Elektromagnetishe Wellen in Drahtes," *Annalen der Physik,* Vol. 30, 1909, p. 905.

[5] Hondros, D., and P. Debye. "Electromagnetishe Wellen in dielektrichen Drahtes," *Annalen der Physik,* Vol. 32, 1910, p. 465.

[6] Zahn, H. "Detection of electromagnetic waves along dielectric wires," *Annalen der Physik,* Vol. 49, 1916, p. 907.

[7] Kiely, D. G. "Dielectric Aerials," *Methuen, Monograph Series,* 1953.

[8] Snitzer, E. "Cylindrical dielectric waveguide modes," *Journal Optical Society of America,* Vol. 5, 1961, pp. 491–498.

[9] Snitzer, E., and H. Osterberg. "Observed dielectric waveguides in the visible spectrum," *Journal Optical Society of America,* Vol. 5, 1961, pp. 499–505.

[10] Kao, K. C., and G. A. Hockham. "Dielectric fiber surface waveguide for optical frequencies," *Proc. IEEE,* Vol. 113, 1966, pp. 1151–1158.

[11] Werts, A. "Propagation de la lumière cohérente dans les fibres optiques," *L'Onde Électrique,* Vol. 46, 1966, pp. 967–980.

[12] Maurer, R. D. "Glass fibers for optical communications," *Proc. IEEE,* Vol. 61, 1973, pp. 452–462.

[13] Gloge, D. "Weakly guiding fibers," *Applied Optics,* Vol. 10, 1971, pp. 2252–2258.

[14] Chu, L. J. "Electromagnetic waves in elliptic hollow pipes of metal," *Journal Applied Physics,* Vol. 9, 1938, pp. 583–591. Note:—A correction to Chu's analysis of one of the higher order modes appeared in 1990 viz. Goldberg, D. H., L. J. Laslet, and R. A. Rimmer. "Modes of elliptical waveguides: a correction," *IEEE Transactions on Microwave Theory and Techniques,* Vol. 38, 1990, pp. 1603–1608.

[15] Lyubimov, L. A., G. I. Veselov, and N. A. Bei. "Dielectric waveguide with elliptical cross-section," *Radio Engineering and Electronics* (USSR), Vol. 6, 1961, pp. 1668–1677.

[16] Yeh, C. "Elliptical dielectric waveguides," *Journal Applied Physics,* Vol. 33, 1962, pp. 3235–3243.

[17] Piefke, G. "Grundlagen sur Berechnung der Ubertragungseigenschafter elliptischer Wellenleiter," *A.E.U.,* Vol. 18, 1964, pp. 4–8.

[18] Dyott, R. B., and J. R. Stern. "Group delay in glass fiber waveguides," *Electronics Letters,* Vol. 7, 1971, pp. 82–84.

[19] Schlosser, W. O. "Delay distortion in weakly guiding optical fibers due to elliptic deformation of the boundary," *Bell System Technical Journal,* Vol. 51, 1972, pp. 487–492.

[20] Ramaswamy, V., W. G. French, and R. D. Standley. "Polarization characteristics of noncircular-core single mode fibers," *Applied Optics,* Vol. 17, 1978, pp. 3014–3017.

[21] Dyott, R. B., J. R. Cozens, and D. G. Morris. "Preservation of polarization in optical fiber waveguides with elliptical cores." *Electronics Letters,* Vol. 15, 1979, pp. 380–382.

[22] Kim, B. Y., J. N. Blake, S. Y. Huang, and H. J. Shaw. "Use of highly elliptical core fibers for two-mode fiber devices," *Optics Letters,* Vol. 12, 1987, p. 729.

Chapter 2
The Dielectric Slab Waveguide

2.1 THE DIELECTRIC SLAB WAVEGUIDE

A dielectric slab of dielectric constant ε_1 with a thickness $2b$ is immersed in a medium ε_2. The slab extends to $\pm$ infinity in the y direction and it is assumed that there can be no variations of field in that direction. Waves propagate in the z direction (see Figure 2.1).

The guide can support two sets of modes, designated E_{m0} and H_{m0} with components of electric and magnetic fields in the direction of propagation. The suffix m denotes the number of periodic variations in the x direction. Lines of E (full) and H (dotted) are shown below for the fundamental E_{10} and H_{10} modes. Maxwell's equations in Cartesian form give the periodic functions as $\sin(u)$, $\cos(u)$ and the evanescent function as $e^{-(w)}$.

Substituting in the boundary-matching equation (1.5) and assuming μ_1 and μ_2 are unity produces the transcendental equations

$$\left(\frac{\sin'(u)}{u \sin(u)} + \frac{e'^{-(w)}}{we^{-(w)}} \right) \left(\varepsilon_1 \frac{\sin'(u)}{u \sin(u)} + \varepsilon_2 \frac{e'^{-(w)}}{we^{-(w)}} \right) = 0 \qquad (2.1)$$

(since n, the variation in the y direction, is, by definition, zero) and

$$\left(\frac{\cos'(u)}{u \cos(u)} + \frac{e'^{-(w)}}{we^{-(w)}} \right) \left(\varepsilon_1 \frac{\cos'(u)}{u \cos(u)} + \varepsilon_2 \frac{e'^{-(w)}}{we^{-(w)}} \right) = 0 \qquad (2.2)$$

The first equation gives the solutions to modes with a field maximum in the center of the slab, which are therefore designated as "even" modes. Equating each bracket to zero gives, for the dielectric-insensitive H modes

---

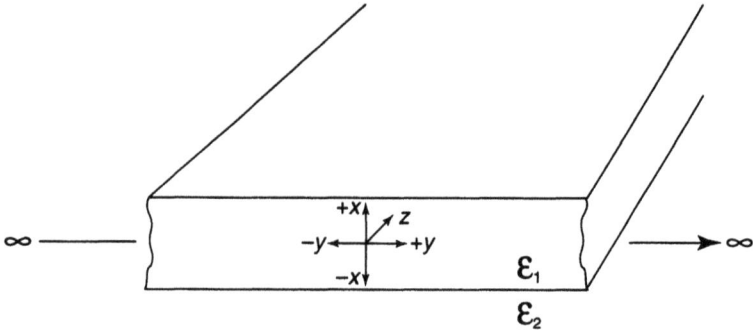

Figure 2.1 Slab waveguide—coordinate system.

$$\tan(u) = \frac{w}{u} \tag{2.3}$$

and for the dielectric-sensitive (charges on the boundary) E modes

$$\tan(u) = \frac{\varepsilon_1}{\varepsilon_2} \frac{w}{u} \tag{2.4}$$

Equation (2.2) gives the solutions to modes with a field minimum in the center of the slab, designated as "odd" modes. For the H modes

$$\tan(u) = -\frac{u}{w} \tag{2.5}$$

and for the E modes

$$\tan(u) = -\frac{\varepsilon_2}{\varepsilon_1} \frac{u}{w} \tag{2.6}$$

Figures 2.2 and 2.3 illustrate the field distribution of the H_{10} and E_{10} modes.

2.2 MODAL CUTOFF

When normalized propagation constant $\bar{\beta} = \beta/k$ becomes equal to the refractive index $n_2 = \sqrt{\varepsilon_2}$, the phase velocity is that of an unguided wave in the cladding and the mode is cut off.

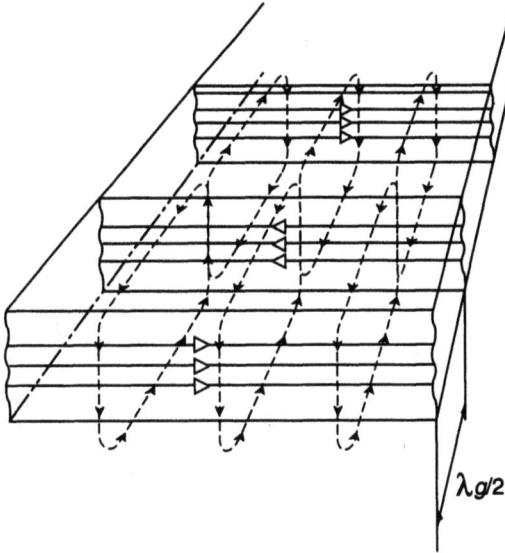

Figure 2.2 H_{10} mode on a dielectric slab: ———, E field; – – – –, H field.

With $\bar{\beta} = n_2$, then $w = 0$ and $u = V_c = V$ at cutoff. Cutoff for the even modes is therefore given by

$$\sin V_c = 0 \tag{2.7}$$

$$V_c = 0, \pi, 2\pi. \ldots \text{ etc.} \tag{2.8}$$

And for the odd modes

$$\cos V_c = 0 \tag{2.9}$$

$$V_c = \frac{\pi}{2}, \frac{3\pi}{2}, \frac{5\pi}{2} \ldots \text{ etc.} \tag{2.10}$$

2.3 PROPAGATION CONSTANTS

Figure 2.4 shows $\bar{\beta}$ against V_b for the fundamental H_{10}, E_{10}, and the first higher order H_{20}, E_{20} modes of a dielectric slab $\varepsilon_1 = 2.25(n_1 = 1.5)$ in free space.

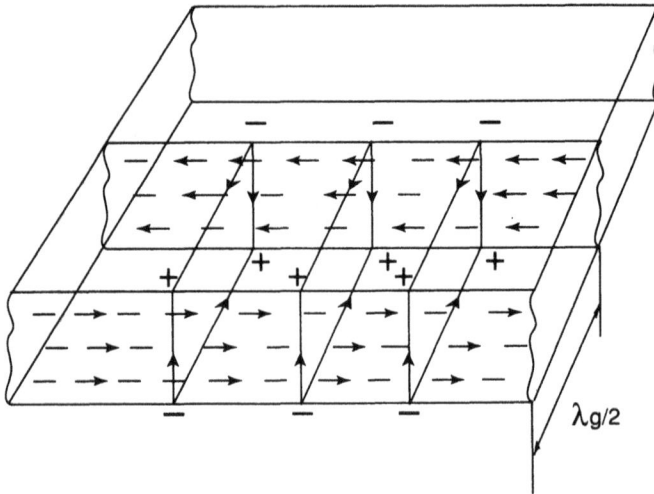

Figure 2.3 E_{10} mode on a dielectric slab. ———, E field; – – – –, H field.

A property which is of great interest in the elliptical waveguide is the difference in normalized propagation constants $\Delta\bar{\beta}$ between the fundamental H_{10} and E_{10} modes—the birefringence of optical fibers.

The value $\Delta\bar{\beta}$ is proportional to the square of the index difference Δn between core and cladding. The relationship holds to within a few percent for most practical fibers. For instance, with $n_1 = 1.5$ and $V_b = 1.0$, then at $\Delta n = 0.01$, $\dfrac{\Delta\bar{\beta}}{(\Delta n)^2} = 0.40$, and at $\Delta n = 0.10$, $\dfrac{\Delta\bar{\beta}}{(\Delta n)^2} = 0.41$.

Figures 2.5 and 2.6 show $\dfrac{\Delta\bar{\beta}}{(\Delta n)^2}$ versus V_b for a dielectric slab, $n_1 = 1.5$ for the H_{10}, E_{10} and H_{20}, E_{20} modes, respectively.

2.4 GROUP VELOCITY

Group velocity, the velocity at which the energy of a wave is propagated, is not an important parameter in present applications of slab waveguides, but it is interesting for showing what happens at the limit of core ellipticity in a fiber.

The phase velocity of a wave is

$$v_p = \frac{\omega}{\beta}$$

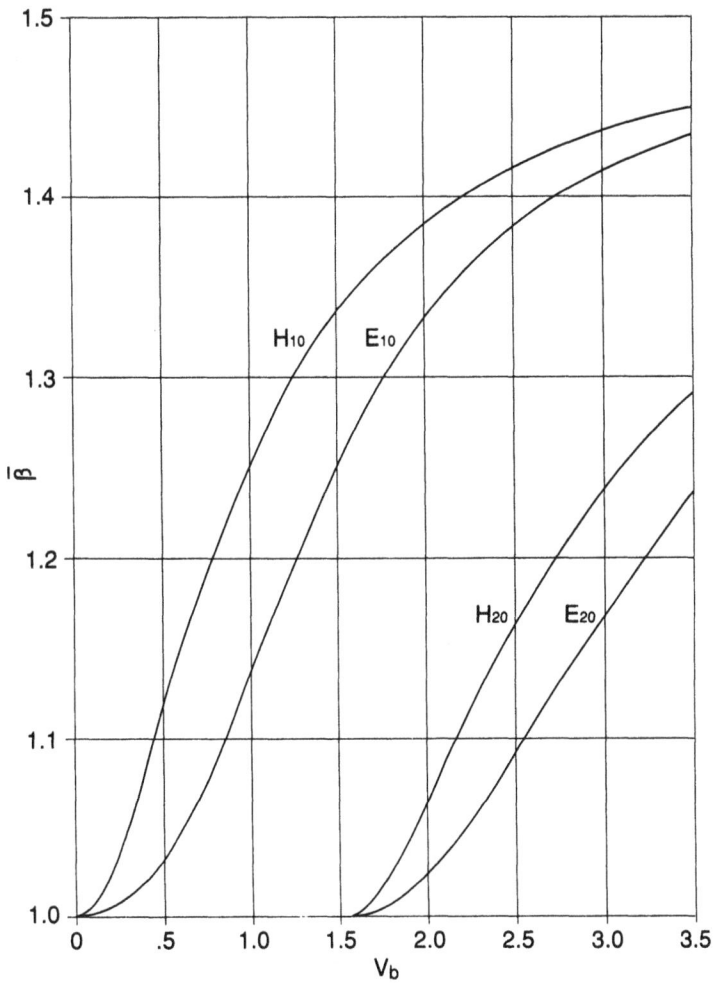

Figure 2.4 Normalized propagation constants for the H_{10}, E_{10}, H_{20}, and E_{20} modes on a dielectric slab.

where ω is the radian frequency $2\pi f$. The group velocity v_g is the slope of the $\omega v\beta$ curve

$$v_g = \frac{\partial \omega}{\partial \beta}$$

Normalizing to the freespace velocity c

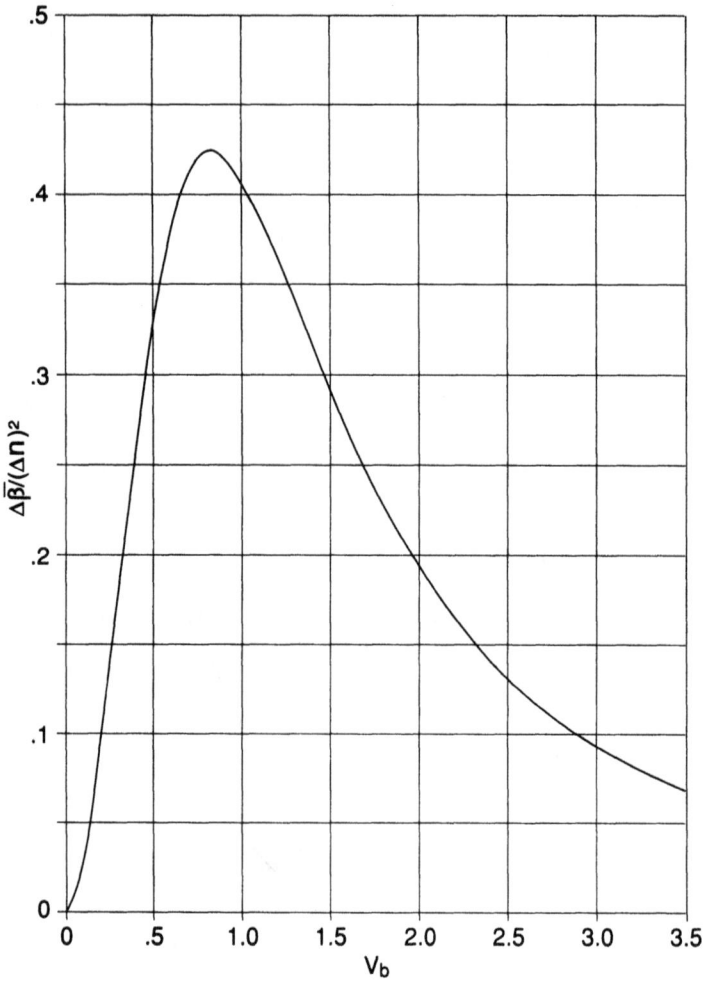

Figure 2.5 Normalized difference in propagation constants for the fundamental H_{10} and E_{10} modes on a dielectric slab.

$$\bar{v}_g = \frac{v_g}{c} = \left[V \frac{\partial \bar{\beta}}{\partial V} + \bar{\beta} \right]^{-1} \tag{2.11}$$

$$\text{with } V = \frac{\omega}{c} b \, (n_1^2 - n_2^2)^{\frac{1}{2}}$$

$$\text{and } \bar{\beta} = \beta \frac{c}{\omega}$$

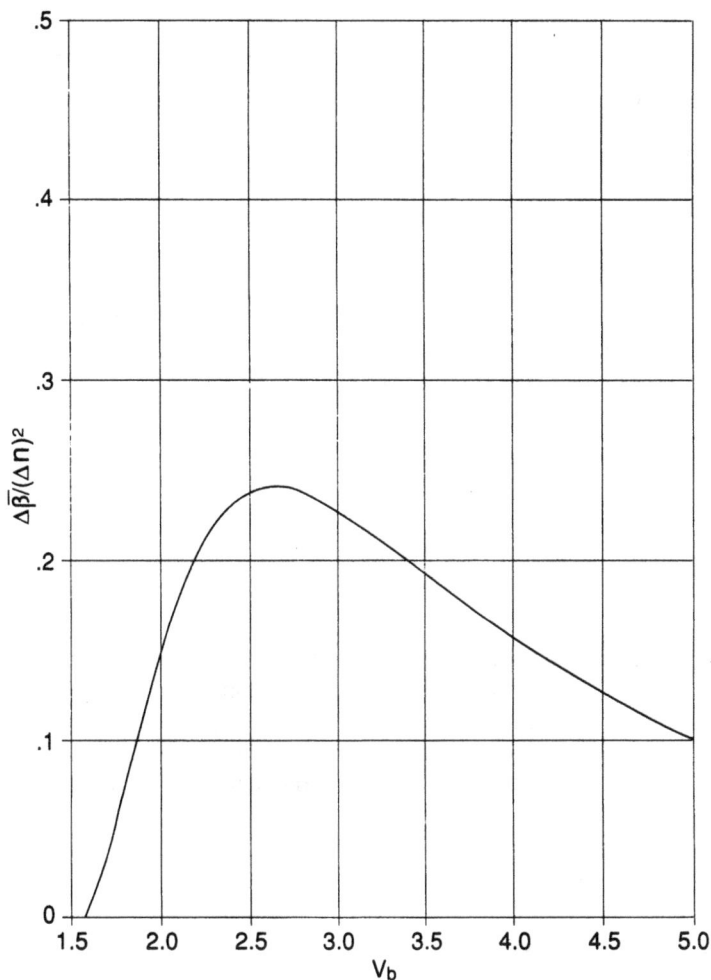

Figure 2.6 Normalized difference in propagation constants for the higher order H_{20} and E_{20} modes on a dielectric slab.

The reciprocal $\dfrac{1}{v_g} = \dfrac{c}{v_g}$ is called the group index n_g. The variable n_g for the fundamental modes in a slab, with $n_1 = 1.5$, $n_2 = 1.0$, is plotted against V_b in Figure 2.7.

The difference in group velocity between the two orthogonally polarized fundamental modes is an important parameter in elliptical core fibers used in interferometric sensors such as the fiber-optic gyro. The variation of group index difference, n_g, of the

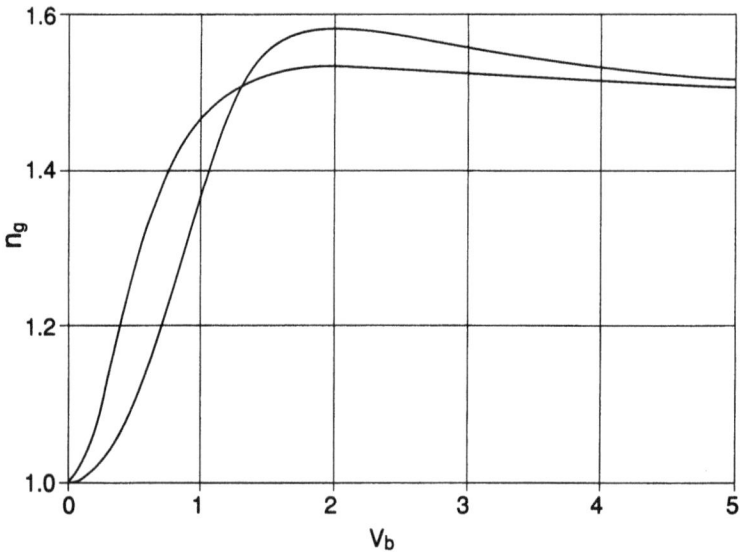

Figure 2.7 Group index for the fundamental H_{10} and E_{10} modes.

fundamental modes is shown in Figure 2.8. At the lower values of V_b, both the phase and group velocities of the E_{10} "fast" mode are greater than those of the H_{10} "slow" mode. However, at higher V values the group velocity of the E_{10} "fast" mode is less than that of the H_{10} "slow" mode, with zero difference at the crossover. Thus, the fast mode defined by phase velocity becomes the slow mode defined by group velocity and vice versa.

2.5 POWER DISTRIBUTION

The fraction of the total power carried in the core is often called the modal confinement and is represented by

$$\frac{P_{core}}{P_{total}} = \eta$$

(or Γ in the theory of semiconductor lasers). For the slab, η for the two fundamental modes is given by straightforward expressions involving u, V, and w.

For the H_{10} index-insensitive mode,

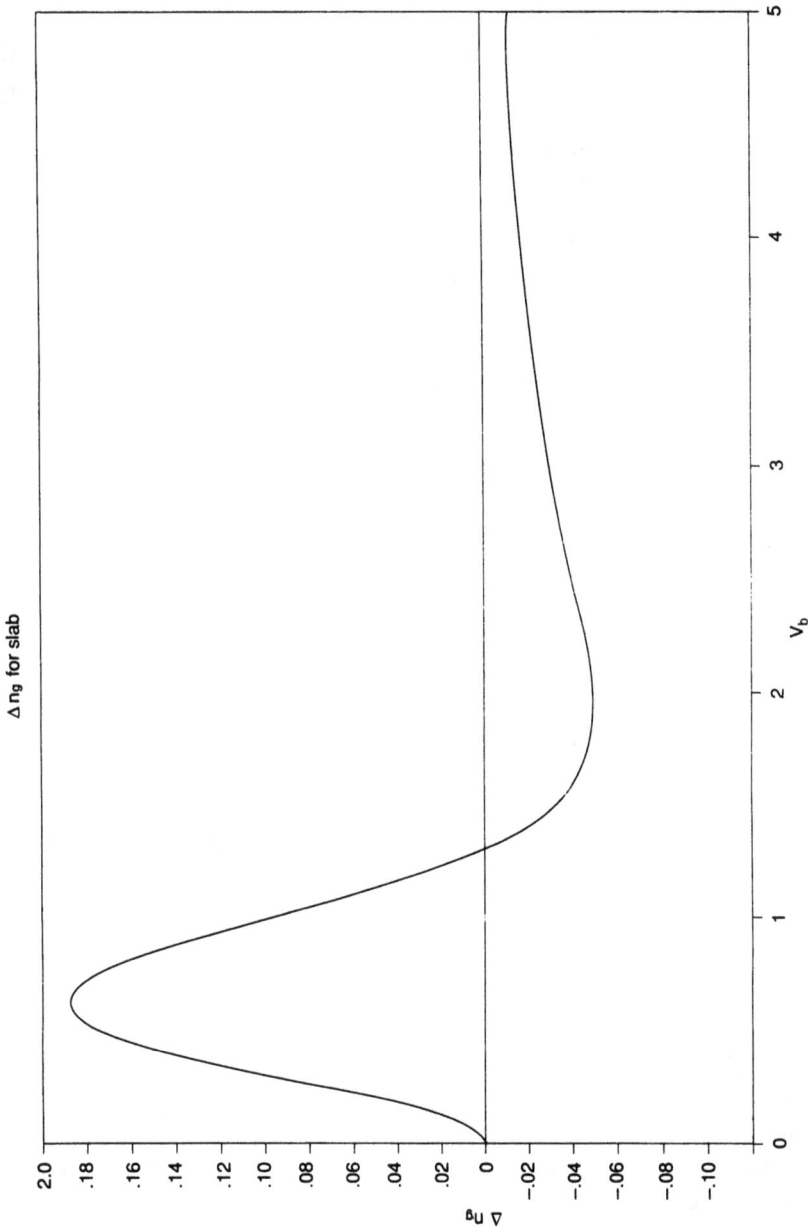

Figure 2.8 Difference in group indices of fundamental H_{10} and E_{10} modes.

$$\eta = 1 - \frac{u^2}{V^2(1 + w)} \tag{2.12}$$

For the E_{10} index-sensitive mode,

$$\eta = 1 - \frac{n_1^2 n_2^2 u^2}{n_1^2 n_2^2 V^2 + n_1^4 w^3 + n_2^4 w u^2} \tag{2.13}$$

For circular and elliptical geometries, the expressions become very complicated and it is easier to use a simple relation that applies to all shapes of waveguides:

$$\frac{c^2}{v_p v_g} = n_1^2 \eta + n_2^2 (1 - \eta) \tag{2.14}$$

(a proof is given by Adams [1]).

Rearranging terms

$$\eta = \frac{\dfrac{\bar{\beta}}{\bar{v}_g} - n_2^2}{n_1^2 - n_2^2} \tag{2.15}$$

For a hollow metal waveguide where there is no power in the cladding, $\eta = 1$, which gives the well-known result

$$\frac{c^2}{v_p v_g} = 1 \tag{2.16}$$

(implying a phase velocity greater than c).

Figure 2.8 shows the power distribution as a function of V_b for the H_{10} and E_{10} modes with $n_1 = 1.5$, $n_2 = 1.0$, and $n_2 = 1.25$. The figure demonstrates the way that the power distribution varies with index difference Δn. For the index-insensitive H_{10} mode, the distribution does not change. For the E_{10} mode, sensitive to Δn, the intuitive feeling is that the larger Δn (and therefore the stronger the guidance), the more the power should be concentrated in the core. However, the reverse is true. For a given V_b, the core width increases as Δn decreases, eventually reaching infinity and therefore carrying all the power as Δn goes to zero.

2.6 FAR-FIELD RADIATION PATTERN

The electric field at a point many wavelengths away from a radiating aperture is the sum, in amplitude and phase, of the fields contributed by each element of the aperture.

For the H_{10} mode, the component of field in the y direction, E_y, varies as a cosine function of the distance x from the center of the slab. Normalized with respect to unity at the core-cladding boundary,

$$E_y = \frac{\cos\left(u\frac{x}{b}\right)}{\cos(u)} \qquad (2.17)$$

Outside the slab, the field dies away to zero at infinity and

$$E_y = \frac{e^{-\left(w\frac{x}{b}\right)}}{e^{-w}} \qquad (2.18)$$

Assuming that these fields are not modified at the end surface of the slab, then, using the Fourier transform to sum over the aperture and substituting $\tan u = \dfrac{w}{u}$, the variation of normalized power distribution P with off-axis angle θ becomes

$$P = \left[\frac{u^2 w^2}{(u^2 - \alpha^2)(w^2 + \alpha^2)}\left(\cos\alpha - \alpha\sin\alpha\,\frac{\cos u}{u\sin u}\right)\right]^2 \qquad (2.19)$$

with

$$\alpha = kb\sin\theta = \frac{V_b}{(n_1^2 - n_2^2)^{\frac{1}{2}}}\sin\theta \qquad (2.20)$$

and

$$u^2 + w^2 = V_b^2$$

The equation has been written in this form for a later comparison with waveguide with circular geometry.

When V_b is large, the aperture is wide in terms of $\dfrac{(n_1^2 - n_2^2)^{\frac{1}{2}}}{\lambda_0}$. The power is almost

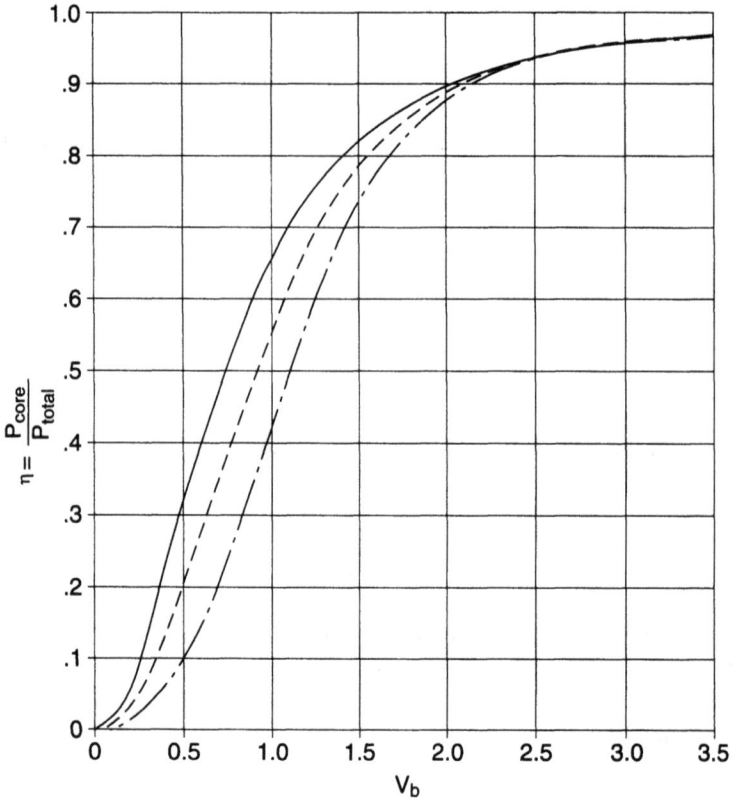

Figure 2.9 Normalized power distribution for the fundamental modes: ————, H_{10} for $n_1 = 1.5$, E_{10} for $\Delta n \rightarrow$ 0; — — —, E_{10} for $n_1 = 1.5$, $\Delta n = 0.25$; — — - — —, E_{10} for $n_1 = 1.5$, $\Delta n = 0.50$.

totally confined to the core ($\eta = 1$) and the evanescent field in the cladding falls off abruptly with distance from the boundary. The radiation pattern of a large aperture having a sharply defined boundary has a main lobe with sidelobes as shown in Figure 2.10(a). As V_b decreases, the aperture narrows and the main lobe widens reaching a maximum. As V_b continues to decrease, the main lobe narrows once more as the field spreads out onto the cladding and the effective aperture enlarges. Now, however, there is no sharp change in field at the boundary and the shape of the field distribution across the whole aperture resembles that of a gaussian. Since the Fourier transform of a gaussian is a gaussian, the sidelobes disappear leaving a pear-shaped main lobe.

Figure 2.10 shows the radiation patterns when:
(a) V_b is large, and the power is mostly confined to the core;
(b) The width of the main lobe, at the half-power points, is a maximum—

$$V_b = \frac{\pi}{2};$$

(c) V_b is small and the field extends far into the cladding.

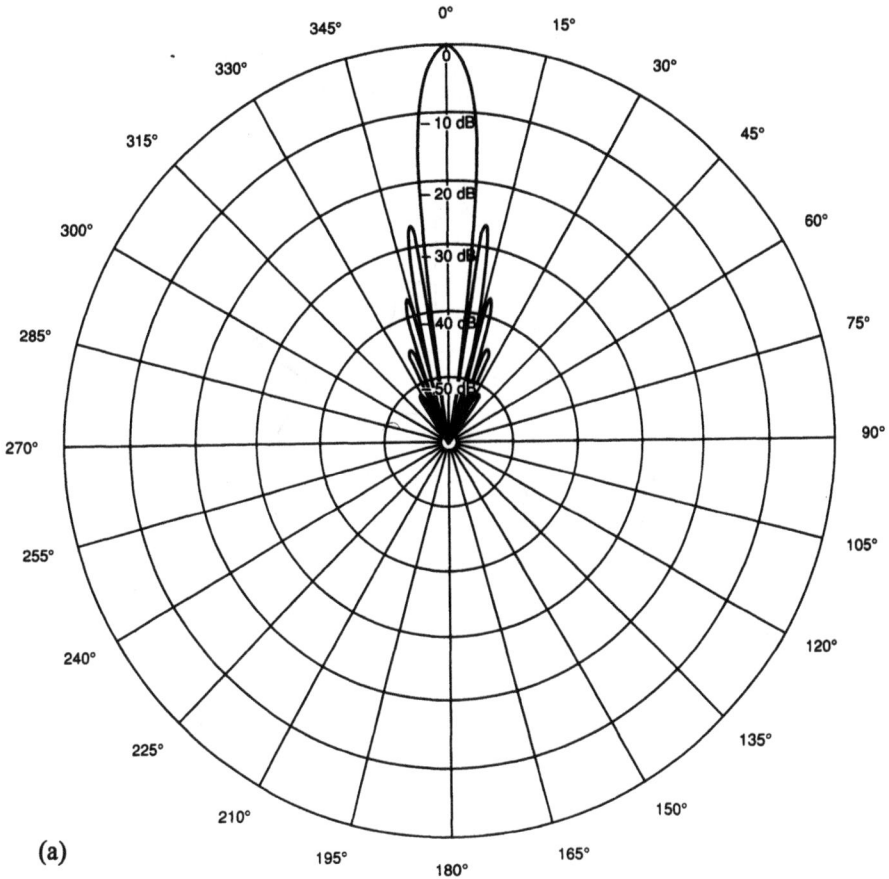

(a)

Figure 2.10 Radiation pattern from slab guide. (a) $n_1 = 1.50$, $n_2 = 1.49$; $V = 5.0$; $u = 1.3059$. (b) $n_1 = 1.50$; $n_2 = 1.49$; $V = \frac{\pi}{2}$ for θ_h maximum; $u = 0.9428$. (c) $n_1 = 1.50$; $n_2 = 1.490$; $V = 0.20$; $u = 0.1962$.

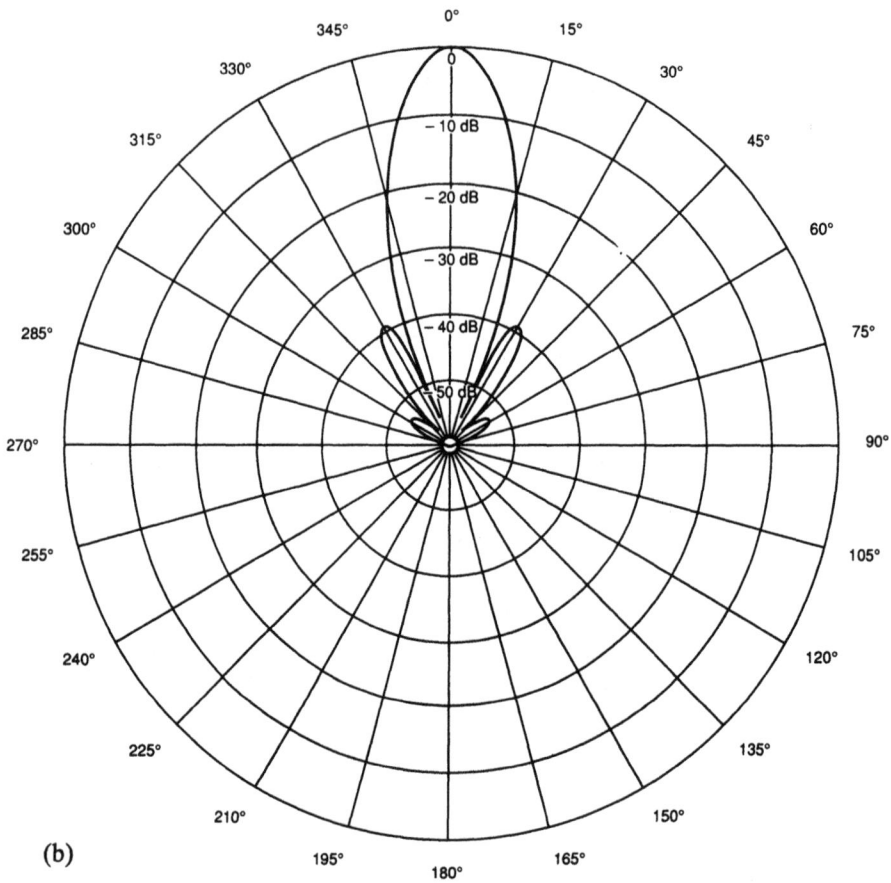

(b)

Figure 2.10 (continued)

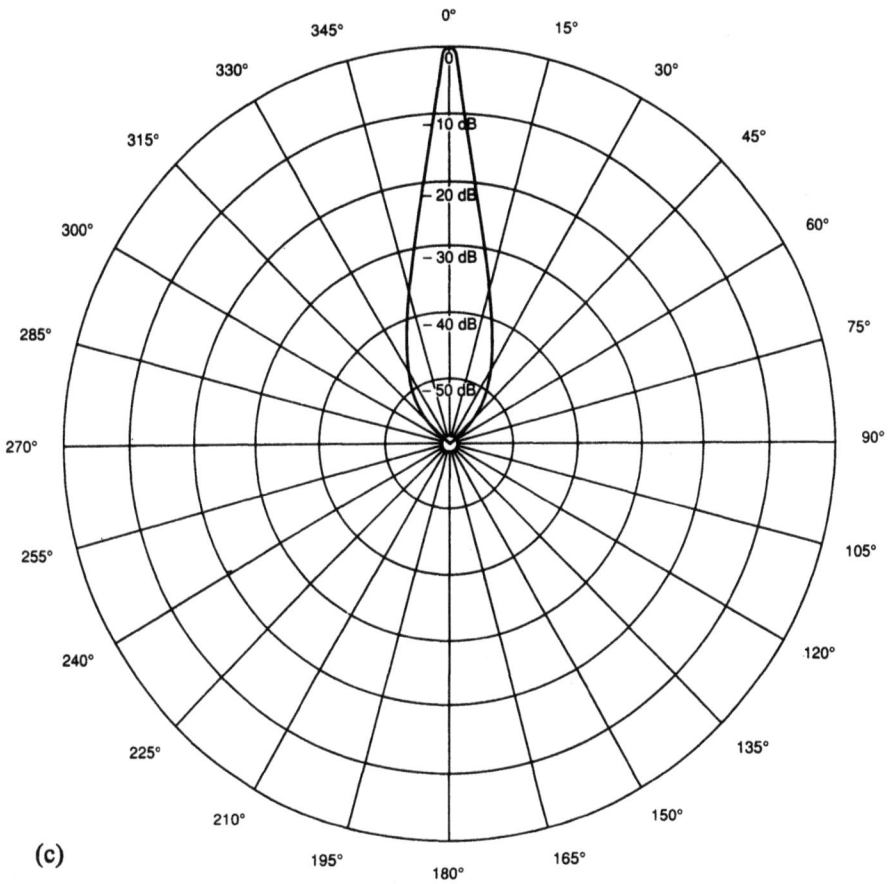

(c)

Figure 2.10 (continued)

REFERENCE

[1] Adams, M. J. *An Introduction to Optical Waveguides,* Wiley, 1981, p. 41.

Chapter 3
Circular Dielectric Waveguide

3.1 CIRCULAR DIELECTRIC WAVEGUIDE

Waves propagate along a cylindrical rod (core) of radius a and dielectric constant ε_1 (refractive index n_1) surrounded by an infinite medium (cladding) of a lower dielectric constant ε_2 (refractive index n_2). The system of coordinates is shown in Figure 3.1.

The variable r is the radial coordinate, ϕ the azimuthal coordinate, and z the axial coordinate in the direction of propagation. The solutions to the wave equations in cylindrical geometry are in terms of Bessel functions. The variable F_{periodic} becomes the quasiperiodic J_n Bessel function. Unlike the plane fields in rectangular coordinates of the infinite slab, the amplitudes of the fields in a rod decrease with radial distance so that the periodic Bessel function J_n resembles a damped sine or cosine function. The variable $F_{\text{evanescent}}$ becomes the quasi-exponential K_n Bessel function. Matching tangential fields at the boundary, (1.5) becomes, assuming μ_1 and μ_2, the relative permeabilities of the core and the cladding are unity,

$$\left(\frac{J'_n(u)}{uJ_n(u)} + \frac{K'_n(w)}{wK_n(w)}\right)\left(\varepsilon_1\frac{J'_n(u)}{uJ_n(u)} + \varepsilon_2\frac{K'_n(w)}{wK_n(w)}\right) = n^2\left(\frac{1}{u^2} + \frac{1}{w^2}\right)\left(\frac{\varepsilon_1}{u^2} + \frac{\varepsilon_2}{w^2}\right) \tag{3.1}$$

Again, the two bracketed terms on the left-hand side represent solutions for modes insensitive to and sensitive to the dielectric constants ε_1 and ε_2 of core and cladding. The variable n is the number of field variations in the azimuthal direction. For $n = 0$ (no azimuthal variation), the modes are decoupled and have either E or H in the direction of propagation. For $n \neq 0$, the modes are coupled by the right-hand side to form the hybrid modes HE and EH. The equations linking u, w, V, and the propagation constant now involve "a," the core radius rather than "b," the slab semithickness

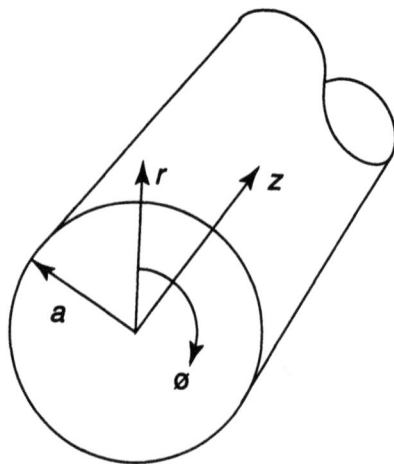

Figure 3.1 Cylindrical coordinate system.

$$u^2 = k_1^2 a^2 - \beta^2 a^2$$

$$w^2 = \beta^2 a^2 - k_2^2 a^2$$

$$V^2 = u^2 + w^2 = k_1^2 a^2 - k_2^2 a^2 \qquad (3.2)$$

$$k_1^2 = \left(\frac{2\pi}{\lambda_0}\right)^2 \varepsilon_1$$

$$k_2^2 = \left(\frac{2\pi}{\lambda_0}\right)^2 \varepsilon_2$$

For modes with no azimuthal variation of field, then $n = 0$ and either

$$\left(\frac{J_n'(u)}{uJ_n(u)} + \frac{K_n'(w)}{wK_n(w)}\right) = 0 \qquad (3.3)$$

which gives $E_z = 0$ for the index-insensitive H_{01} modes, or

$$\left(\varepsilon_1\frac{J_n'(u)}{uJ_n(u)} + \varepsilon_2\frac{K_n'(w)}{wK_n(w)}\right) = 0 \qquad (3.4)$$

which gives $H_z = 0$ for the index-sensitive E_{01} modes.

For $n > 0$ and $\varepsilon_1 \approx \varepsilon_2$ (small difference in dielectric constants)

$$\left(\frac{J_n'(u)}{uJ_n(u)} + \frac{K_n'(w)}{wK_n(w)}\right) = n\left(\frac{1}{u^2} + \frac{1}{w^2}\right) \tag{3.5}$$

Using the differential formulas for Bessel functions

$$J_n'(u) = -\frac{nJ_n(u)}{u} + J_{n-1}(u) = \frac{nJ_n(u)}{u} - J_{n+1}(u) \tag{3.6}$$

$$K_n'(w) = -\frac{nK_n(w)}{w} - K_{n-1}(w) = \frac{nK_n(w)}{w} - K_{n+1}(w) \tag{3.7}$$

leads to

$$\left(\frac{J_{n-1}'(u)}{uJ_n(u)} + \frac{K_{n+1}'(w)}{wK_n(w)}\right) = 0 \tag{3.8}$$

$$\left(\frac{J_{n+1}'(u)}{uJ_n(u)} + \frac{K_{n-1}'(w)}{wK_n(w)}\right) = 0 \tag{3.9}$$

The former equation belongs to the hybrid EH modes where the axial electric field is greater than the axial magnetic field, and the latter to the hybrid HE modes where the reverse is true.

The field patterns for the HE_{11} mode are shown in Figure 3.2. In Figure 3.3, a comparison is made between metal-walled and dielectric rod waveguides showing the transverse electric fields together with the $\frac{\omega}{\beta}$ phase diagrams.

Because, unlike the H_{11} in metal guide, the fields have no zero except at infinity, the HE_{11} mode has no cutoff frequency (V) below which it will not propagate, so that cutoff is technically at $V = 0$. However, practical waveguides lose guidance as V decreases and the power shifts from the guiding core into the nonguiding cladding. For a dielectric rod in free space with $\varepsilon_1 = 2.3$ and $\varepsilon_2 = 1$, it is difficult to keep guidance when $V < 0.9$ or even to get a computer solution to (3.1) at $V < 0.8$. Figure 3.4 shows the minimum value of V

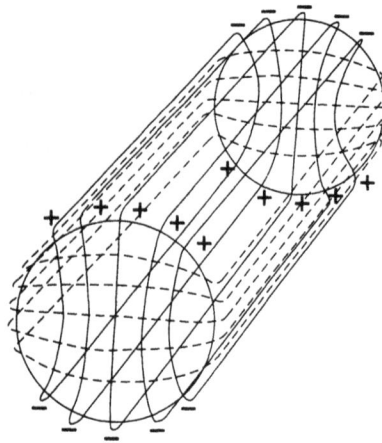

Figure 3.2 Field patterns for the HE_{11} mode. ———, Electric lines; - - - -, magnetic lines.

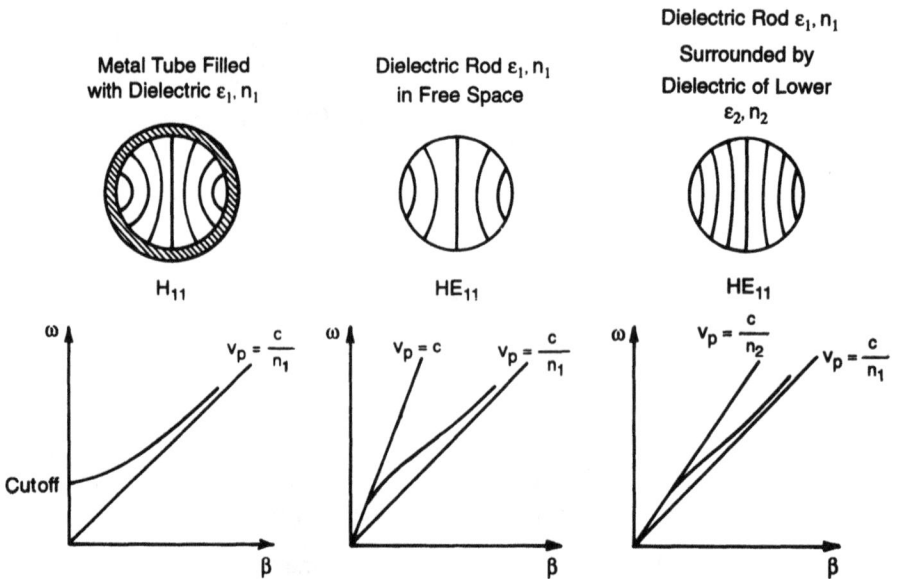

Figure 3.3 Metal and dielectric waveguide.

for a solution for differing $\delta = 1 - \dfrac{\varepsilon_2}{\varepsilon_1}$ when the Bessel functions are accurate to 11 decimal places. The effective bandwidth between minimum V and the higher mode cutoff at $V = 2.405$ diminishes as δ increases.

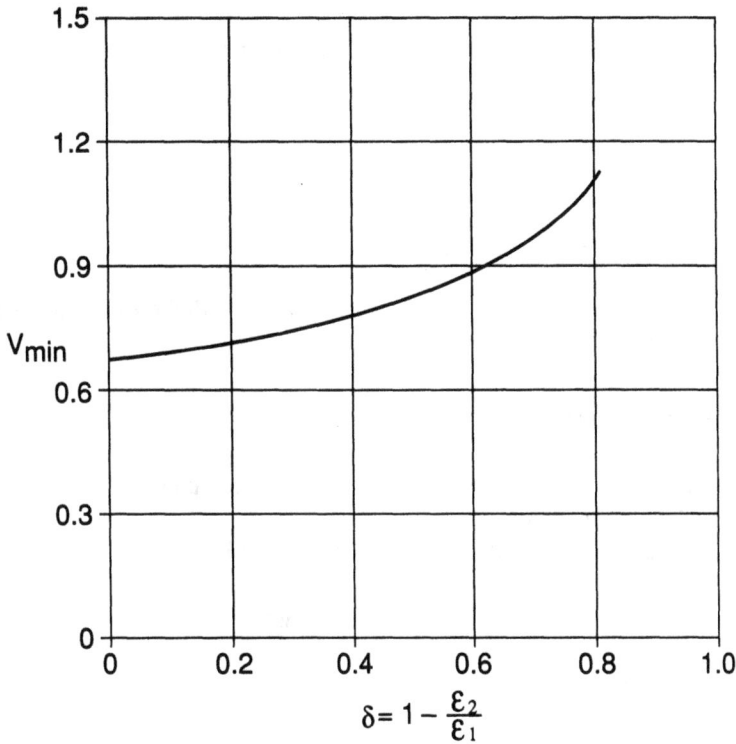

Figure 3.4 Minimum value of V for computer solution to (3.1).

At low values of V and for small δ, an asymptotic expression may be used for w

$$w_{as} = 2 \exp - \left(\varsigma + \frac{J_n(V)}{VJ_n(V)} \right) \tag{3.10}$$

where ς is Euler's constant = 0.577215565.

The accuracy of this approximation for $\delta = 0.01$ and $n_1 = 1.5$ is listed in Table 3.1. A useful approximation for u, particularly for small δ, is given by Gloge [1]:

$$u = \frac{\left(1 + 2^{\frac{1}{2}}\right)V}{1 + \left(4 + V^4\right)^{\frac{1}{4}}} \tag{3.11}$$

Table 3.1
Asymptotic Values of w_{as} for $n_1 = 1.5$, $\delta = 0.01$

V	w_{as}	% Error
0.9	0.12315456	−0.024
1.0	0.19731495	−1.628
1.1	0.27990085	−3.734

Figure 3.5 compares u calculated by this approximation with the exact values from (3.1) for various values of δ. The fit at $\delta = 0.1$ is almost exact.

3.2 MODAL CUTOFF

Cutoff occurs when $w = 0$, and $u = V$. The wave propagates at the velocity of an unguided wave in the cladding medium.

For the E_{0m} and H_{0m} modes, inverting (3.3) and (3.4) gives

$$\frac{uJ_0(u)}{J_0'(u)} = -\frac{wK_0(w)}{K_0'(w)} \tag{3.12}$$

$$\frac{uJ_0(u)}{J_0'(u)} = -\frac{\varepsilon_1}{\varepsilon_2}\frac{wK_0(w)}{K_0'(w)} \tag{3.13}$$

For $w = 0$, the right-hand sides are zero and the cutoff condition for both sets of modes is

$$J_0(u) = J_0(V_c) = 0 \tag{3.14}$$

where V_c is the value of V at cutoff. The cutoff for modes with no azimuthal variation of field is independent of ε_1 and ε_2.

The E_{0m} modes in hollow metal guides have the same cutoff as their counterparts in dielectric guide. However, for the H_{0m} modes there is an extra zero of the tangential electric field at the metal boundary so that the cutoff condition for the H_{01} mode in metal guide becomes

$$J_1(V_c) = 0 \tag{3.15}$$

Returning to dielectric guide for modes with azimuthal variation of field $n > 1$, then as $w \to 0$,

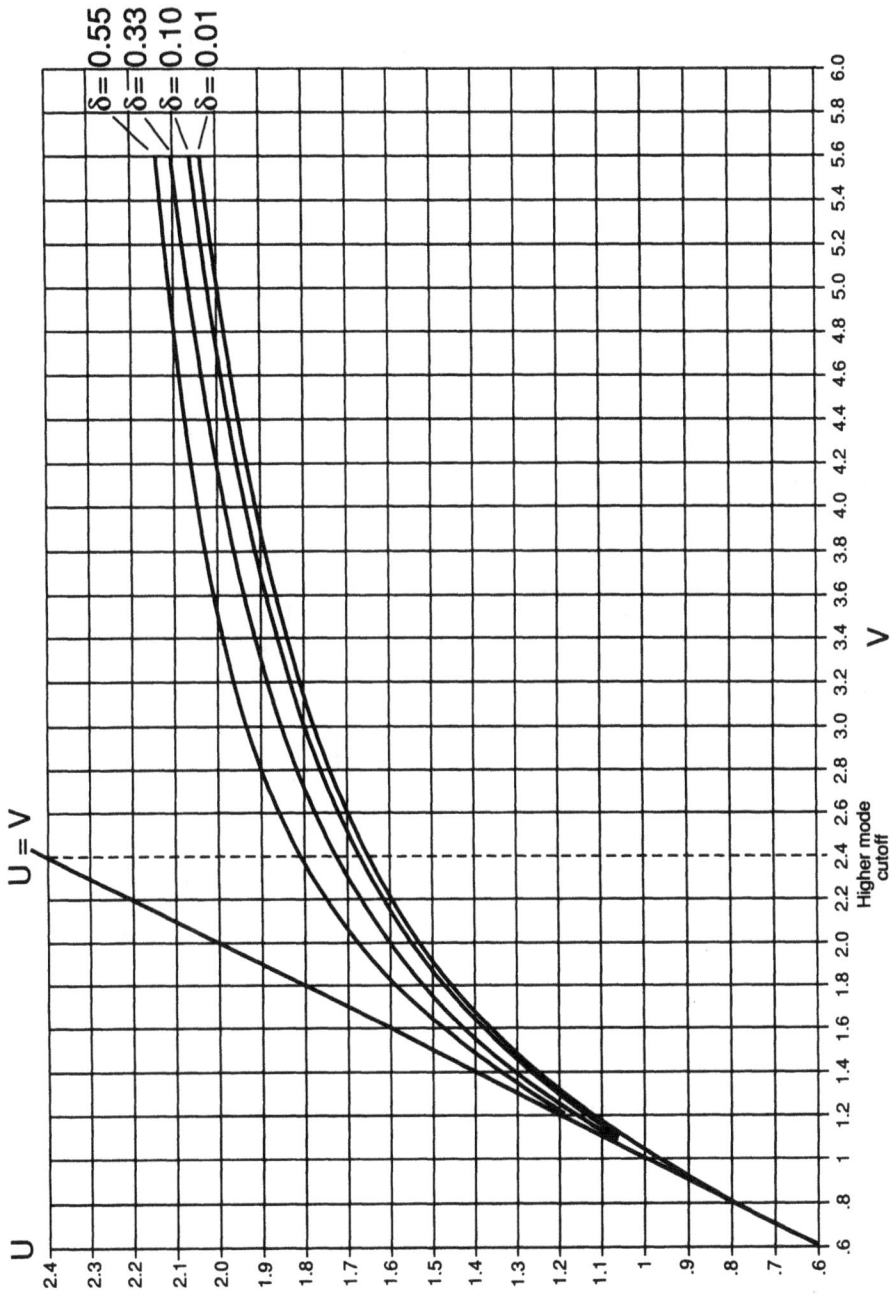

Figure 3.5 Computed values of u from (3.1).

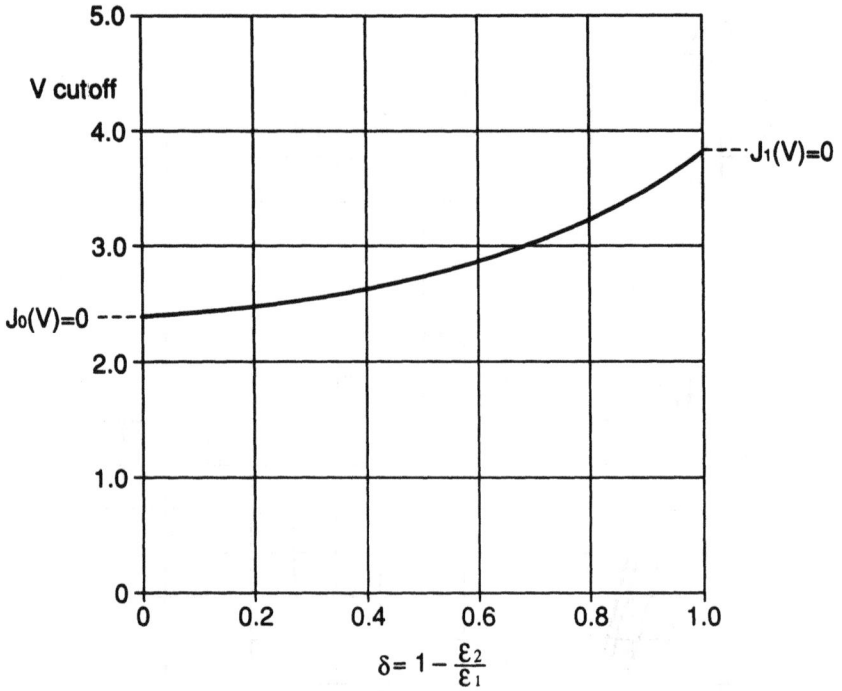

Figure 3.6 Cutoff of the HE_{21} mode as a function of $\delta = 1 - \dfrac{\varepsilon_2}{\varepsilon_1}$.

$$K_n(w) \to (n - 1)! \frac{2^{n-1}}{w^n} \tag{3.16}$$

Substituting in (3.1) leads to the cutoff condition

$$\frac{J_{n-1}(u)}{uJ_n(u)} = \frac{\varepsilon_2}{(n - 1)(\varepsilon_1 + \varepsilon_2)} \tag{3.17}$$

For small dielectric differences $\varepsilon_1 \approx \varepsilon_2$, this reduces to

$$\frac{J_{n-1}(u)}{uJ_n(u)} = \frac{1}{2(n - 1)} \tag{3.18}$$

Figure 3.6 shows how the cutoff of the HE_{21} mode varies with δ. For $\delta \to 0$ as for the "weak guiding" of the fibers used for optical communications, $V_c = 2.405$ and the cutoff

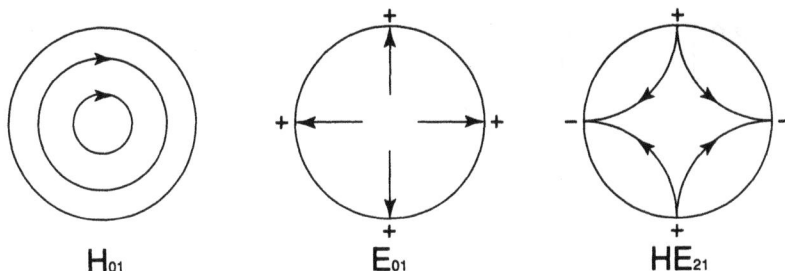

Figure 3.7 Transverse electric field patterns for the first three higher order modes.

is virtually the same as those for the E_{01} and H_{01} modes. V_c increases to 3.832; $J_1(V_c) = 0$ as $\delta \to 1$.

Figure 3.7 shows the patterns of transverse electric field across the core for the first three higher order modes; H_{01}, E_{01} and HE_{21}.

3.3 GROUP VELOCITY

The group velocity v_g is again the slope of the $\dfrac{\omega}{\beta}$ curve (see, for example, Figure 3.3). The group index c/v_g for a circular guide with $n_1 = 1.5$ and $n_2 = 1.0$ is shown in Figure 3.8. As with the slab, the group index exceeds the core index over a range of values for V, giving a group velocity slower than that for an unbounded wave in core material. The phase velocity, however, always lies between the limits of that of an unbounded wave in core or cladding.

3.4 POWER DISTRIBUTION

As mentioned in Chapter 1, an exact explicit expression for $\dfrac{P_{core}}{P_{total}} = \eta$ for the HE_{11} mode in circular dielectric guide is complicated [2]. Because it is generally necessary to solve the transcendental equation by computer to get the values of u and w, it is easier to solve twice, compute the group velocity, and then use the general relationship (2.14) to find η. Figure 3.9 shows the power distribution and the way it varies with index difference for a guide with a core index of 1.50. For small index difference

$$\eta \approx 1 - \frac{u^2}{V^2}\left(1 - \frac{K_0^2(w)}{K_1^2(w)}\right) \tag{3.19}$$

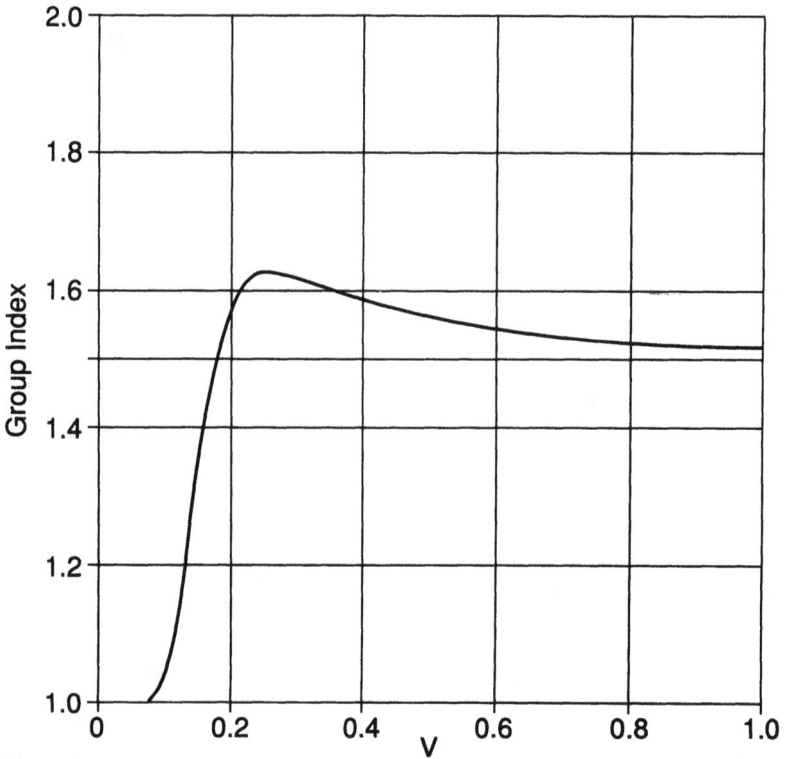

Figure 3.8 Group index of circular guide: $n_1 = 1.50$; $n_2 = 1.0$.

3.5 FAR-FIELD RADIATION PATTERN

Because there is not such a simple solution to the transcendental equation in circular guide as there is in the slab, the analysis of the radiation from a dielectric rod is more involved [3]. A much simplified analysis for small index difference gives, for the normalized power distribution [4]

$$P = \left[\frac{u^2 w^2}{(u^2 - \alpha^2)(w^2 + \alpha^2)} \left(J_0(\alpha) - \alpha J_1(\alpha) \frac{J_0(u)}{u J_1(u)} \right) \right]^2 \tag{3.20}$$

$$\alpha = ka \sin \theta = \frac{V}{\left(n_1^2 - n_2^2 \right)^{\frac{1}{2}}} \sin \theta \tag{3.21}$$

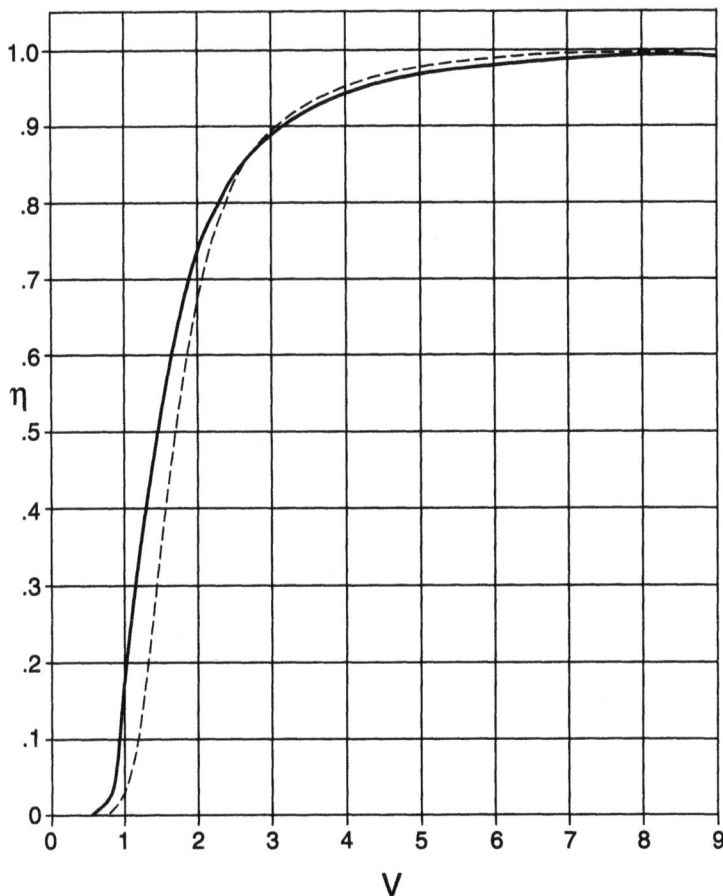

Figure 3.9 Power distribution of the HE_{11} mode: $n_1 = 1.50$; —— $n_2 = 1.49$; - - - - $n_2 = 1.0$.

Comparison with the equivalent equation for the slab (2.19) shows that the cosine function becomes the cosine-like J_0 function and sine the sine-like J_1. The radiation patterns therefore resemble those of the slab with a narrow main lobe together with sidelobes at large V and a narrow main lobe without sidelobes at small V. The maximum half-power angle θ_h of the main lobe occurs at $V = 2.55$, near to but not precisely at higher mode cutoff, as is so with the slab.

Figure 3.10(a,b,c) shows the radiation patterns for small V, large V, and at maximum θ_h.

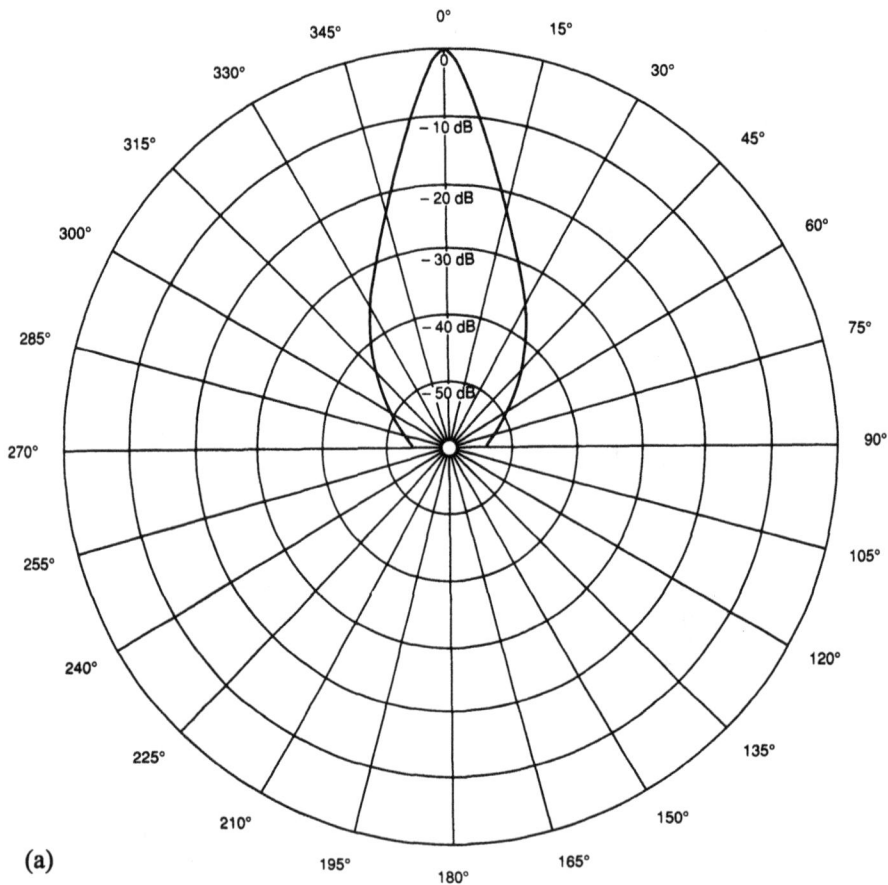

Figure 3.10 Radiation pattern from circular dielectric guide: (a) $n_1 = 1.5$; $n_2 = 1.45$; $V = 1.0$. (b) $n_1 = 1.5$; $n_2 = 1.45$; $V = 2.55$ at θ_h maximum. (c) $n_1 = 1.5$; $n_2 = 1.45$; $V = 4.5$.

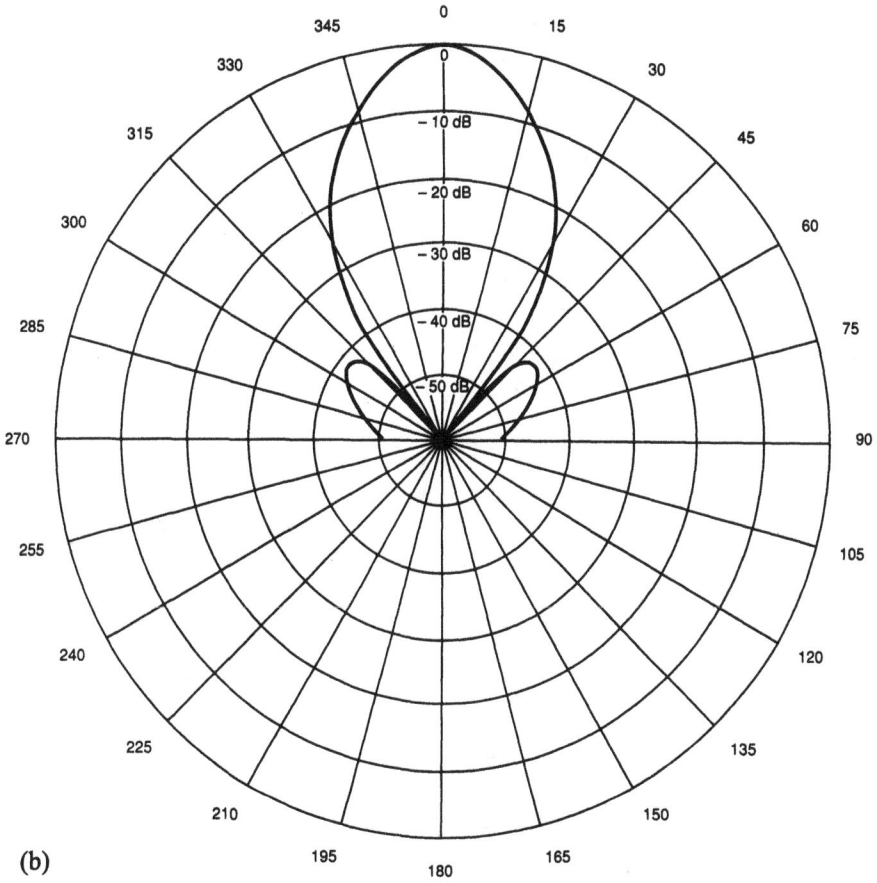

(b)

Figure 3.10 (continued)

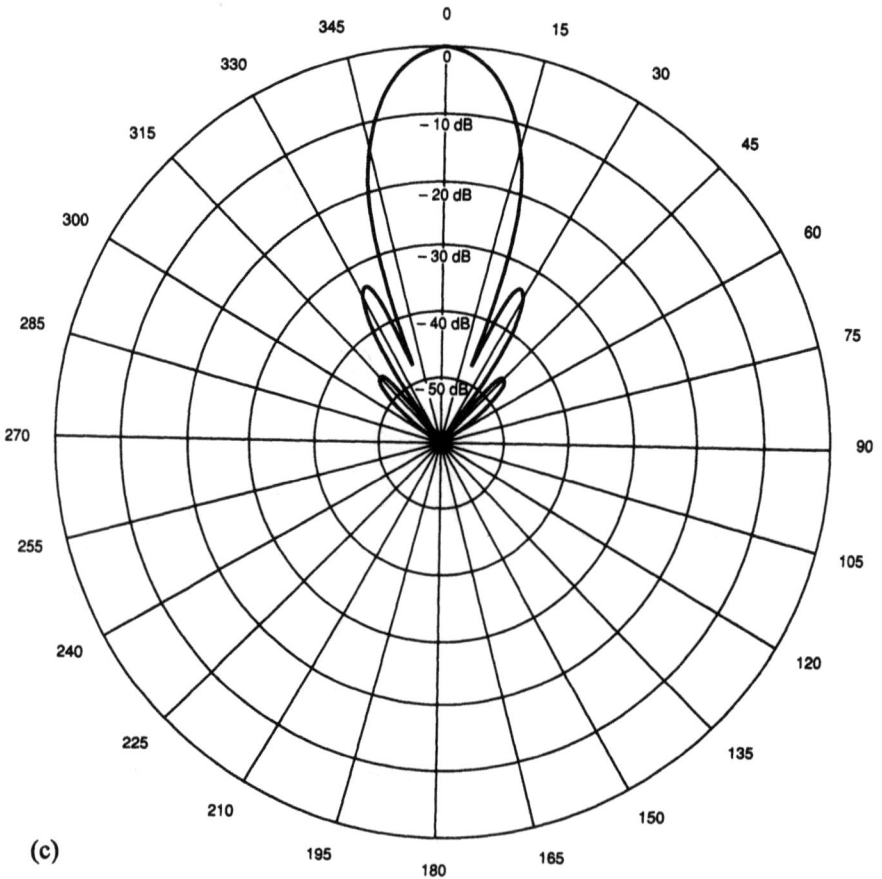

(c)

Figure 3.10 (continued)

REFERENCES

[1] Gloge, D., "Weakly guiding fibers." *Applied Optics*, Vol. 10, 1971, pp. 2252–2258.

[2] Snyder, A. W., and J. D. Love. *Optical Waveguide Theory*, Chapman and Hall: London and New York, 1983, p. 257.

[3] James, J. R. "Theoretical investigation of cylindrical dielectric rod antennas," *Proc. IEE*, Vol. 114, 1967, pp. 309–319.

[4] Gambling, W. A., D. H. Payne, H. Matsumura, and R. B. Dyott. "Determination of core diameter and refractive index difference of single-mode fibers by observation of the far field pattern," *Microwaves, Optics and Acoustics*, Vol. 1, 1976, pp. 13–17.

Chapter 4

Elliptical Dielectric Waveguides

4.1 ELLIPTICAL DIELECTRIC WAVEGUIDES

Waves propagate on a core of elliptical cross section dielectric constant ε_1 (refractive index n_1) surrounded by an infinite medium (cladding) of dielectric constant ε_2 (refractive index n_2). The radial ξ coordinate describes a set of confocal ellipses and the η azimuth coordinate a set of hyperbolae orthogonal to the ellipses (Figure 4.1). The axial coordinate, in the direction of propagation, is z. The elliptical coordinates ξ, η, and z relate to the Cartesian coordinates x, y, and z as

$$x = q \cosh \xi \cos \eta \tag{4.1}$$

$$y = q \sinh \xi \sin \eta \tag{4.2}$$

$$z = z \tag{4.3}$$

Defining the core boundary by $\xi = \xi_0$, the semimajor and semiminor axes of the core ellipse are $a = q \cosh \xi_0$; $b = q \sinh \xi_0$.
The eccentricity e is

$$e = \left[1 - \left(\frac{b}{a} \right)^2 \right]^{\frac{1}{2}} = \frac{1}{\cosh \xi_0} \tag{4.4}$$

The wave equations in elliptical coordinates are

$$\frac{\partial^2 E_z}{\partial \xi^2} + \frac{\partial^2 E_z}{\partial \eta^2} + [q^2 (\varepsilon k_0^2 - \beta^2)(\sinh^2 \xi + \sin^2 \eta)] E_z = 0 \tag{4.5}$$

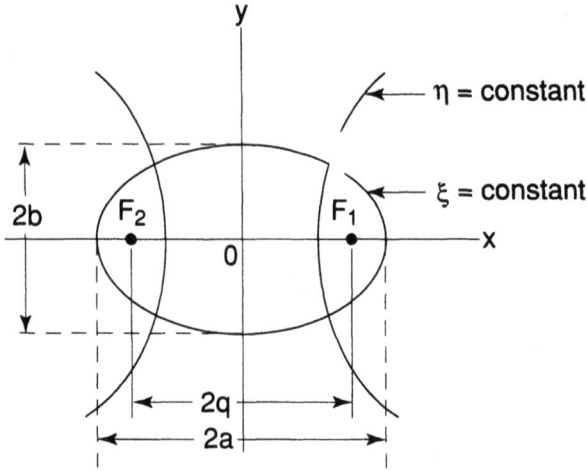

Figure 4.1 Elliptical coordinate system.

$$\frac{\partial^2 H_z}{\partial \xi^2} + \frac{\partial^2 H_z}{\partial \eta^2} + [q^2 (\varepsilon k_0^2 - \beta^2)(\sinh^2 \xi + \sin^2 \eta)] H_z = 0 \tag{4.6}$$

or

$$\frac{\partial^2 \Phi}{\partial \xi^2} + \frac{\partial^2 \Phi}{\partial \eta^2} + [q^2 (\varepsilon k_0^2 - \beta^2)(\sinh^2 \xi + \sin^2 \eta)] \Phi = 0 \tag{4.7}$$

where the dielectric constant $\varepsilon = \varepsilon_1$ in the core ($\xi < \xi_0$) and $\varepsilon = \varepsilon_2$ in the cladding ($\xi > \xi_0$),

$$k_0 = \text{free-space wave number} = \frac{2\pi}{\lambda_0}$$

$$\lambda_0 = \text{free-space wavelength}$$

$$\beta = \text{propagation constant}$$

Separating variables with $\Phi (\xi, \eta) = \Theta(\eta) R(\xi)$

$$\frac{\partial^2 \Theta}{\partial \eta^2} + \left[c - (\varepsilon k_0^2 - \beta^2) \frac{q^2}{2} \cos 2\eta \right] \Theta = 0 \tag{4.8}$$

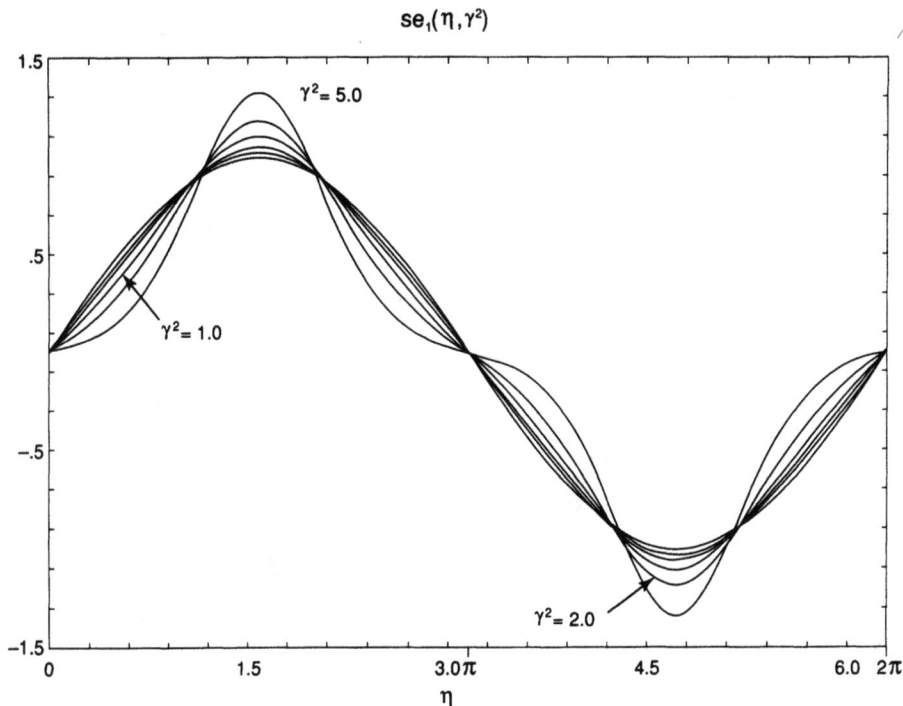

Figure 4.2(a) Mathieu azimuthal function se_1.

$$\frac{\partial^2 R}{\partial \xi^2} - \left[c - (\epsilon k_0^2 - \beta^2) \frac{q^2}{2} \cosh 2\xi \right] R = 0 \qquad (4.9)$$

where c is the separation constant. Equations (4.8) and (4.9) are the Mathieu and modified Mathieu equations.

The solutions to the Mathieu equation (4.8) are in terms of functions that describe the azimuthal variation of field around the ellipse, equivalent to $\cos n\phi$ and $\sin n\phi$ in the circular guide. These functions (the s_e sine-like elliptical and c_e cosine-like elliptical; see Figures 4.2(a,b)) give

$$\Theta(\eta) = \begin{bmatrix} s_{en}\ (\eta,\gamma^2) \\ c_{en}\ (\eta,\gamma^2) \end{bmatrix} \qquad (4.10)$$

$$ce_1(\eta, \gamma^2)$$

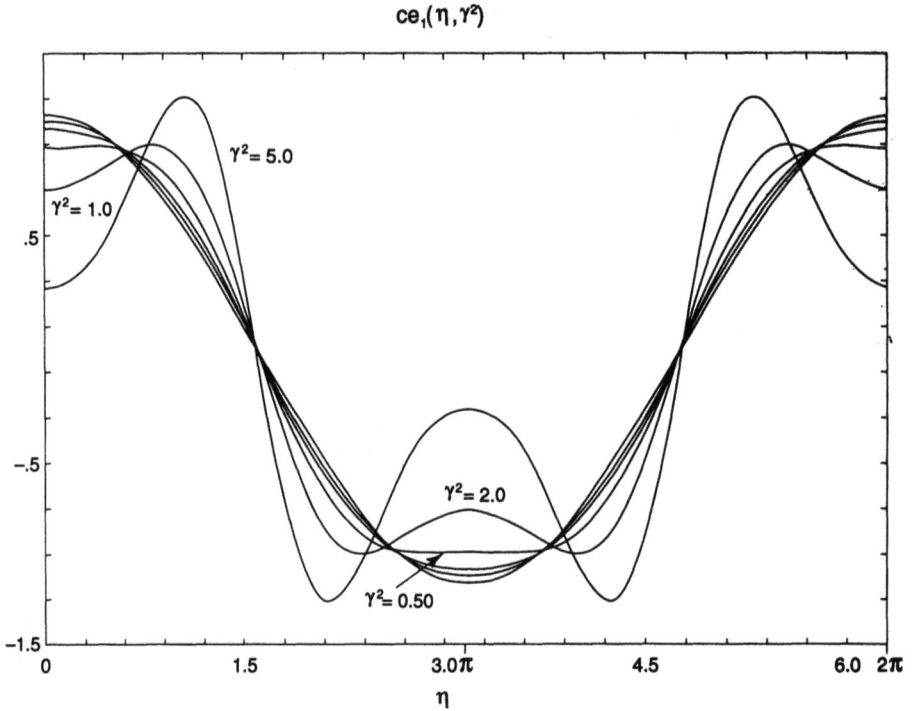

Figure 4.2(b) Mathieu azimuthal function ce_1.

with

$$\gamma^2 = (\varepsilon k_0^2 - \beta^2) \frac{q^2}{4} \tag{4.11}$$

Substituting for q in terms of a and b,

$$q^2 = a^2 - b^2 \tag{4.12}$$

$$\gamma^2 = \left[\left(\frac{a}{b} \right)^2 - 1 \right] \frac{b^2}{4} (\varepsilon k_0^2 - \beta^2) = \left[\frac{a^2}{b^2} - 1 \right] \frac{u^2}{4} \tag{4.13}$$

The term $b^2(\varepsilon k_0^2 - \beta^2)$ corresponds to the function u in circular guide but is related to the semiminor axis b instead of radius a.

For a mode to exist on the dielectric rod it must have a field which is azimuthally periodic; the field pattern must repeat exactly around the perimeter. This in turn means that in order for (4.8) to give a periodic solution, the separation constant c has to take on

$J_1(\xi)$ and $Se_1(\xi,\gamma^2)$ for γ^2= .01, .05, .10, .15, .25, .50, 1.0, 2.0

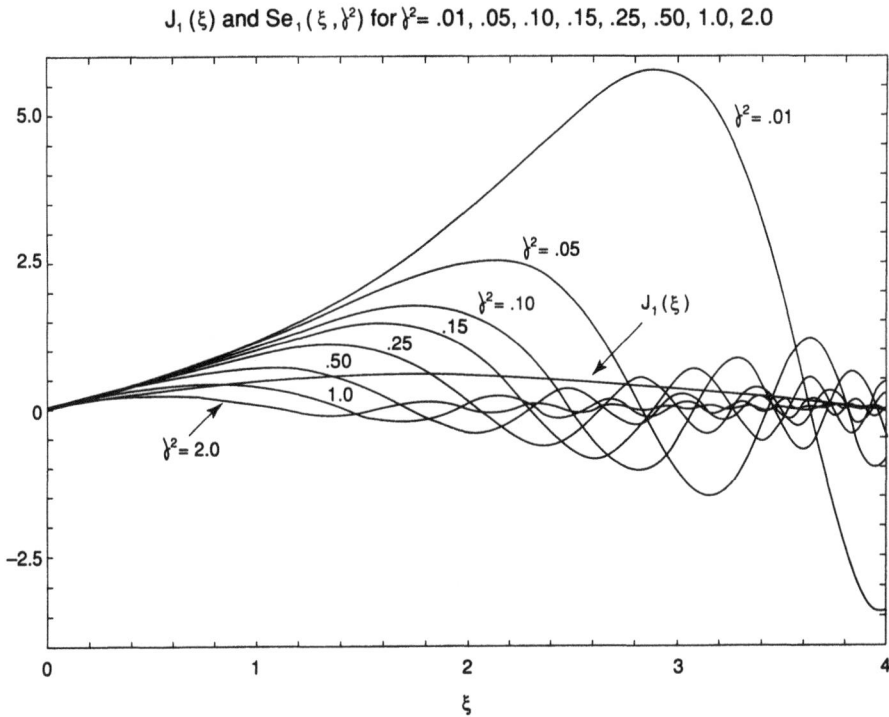

Figure 4.2(c) Mathieu periodic radial function Se_1.

a particular value for each value of γ^2. When γ^2 is not zero, there is a value of c that gives only one even (cosine) or odd (sine) periodic solution to (4.8). These characteristic values of c are denoted by Yeh [1] as $a_n(\gamma^2)$ (even) and $b_n(\gamma^2)$ (odd).

Note that for a circular guide, $q = 0$ and therefore γ^2 is always zero and the solution for (4.8) is in terms of sine and cosine functions; that is, the fields vary in azimuth θ as $\cos n\theta$ and $\sin n\theta$.

The solutions to the modified Mathieu equation (4.9) are in terms of the functions that describe the radial variation of field: These are the periodic S_e (sine-like) and C_e (cosine-like) functions (akin to the Bessel J function) linked to the fields inside the core and the Fek and Gek evanescent functions (akin to the Bessel K functions) outside the core. The nomenclature is that of McLachlan [2]; the k signifying the resemblance to the Bessel K, the e the elliptical subscript, and with F and G chosen as arbitrary letters. There are two types of both periodic and evanescent functions that coalesce to the Bessel J and K functions for the circular core. Figures (4.2(a–g)) show typical s_e, c_e, S_e, C_e, Fek, and Gek functions. Programs for computing these functions are described by Rengarajan and Lewis [3] and Toyama and Shogen [4].

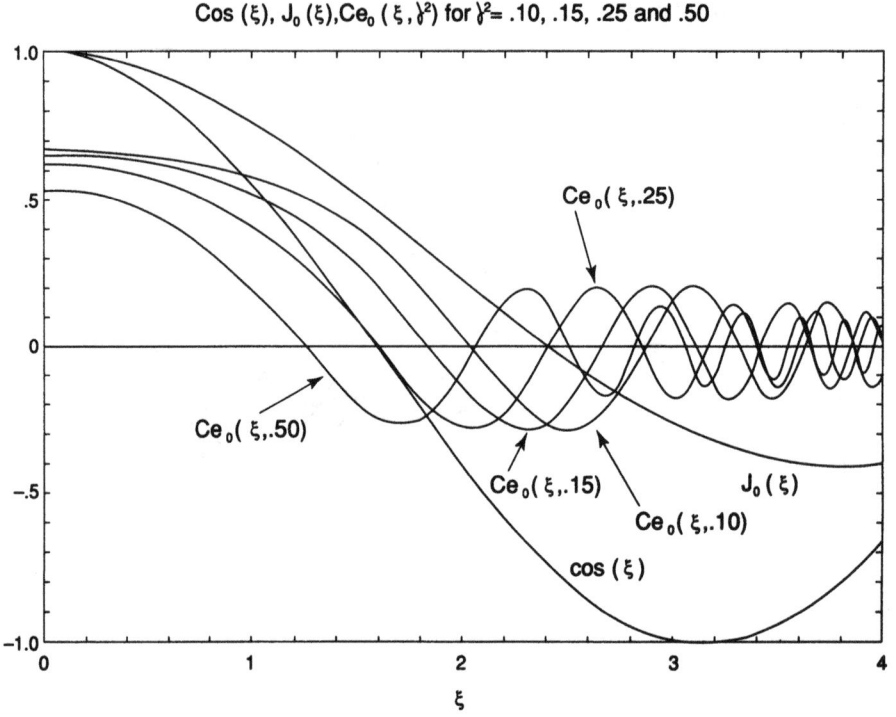

Figure 4.2(d) Mathieu periodic radial function Ce_0.

In order to satisfy the conditions for propagating waves, the tangential electric and magnetic fields must be matched at the core-cladding boundary $\xi = \xi_0$. The fields to be matched are

$$E_{z\ core} = E_{z\ clad} \tag{4.14}$$

$$H_{z\ core} = H_{z\ clad} \tag{4.15}$$

$$E_{\eta\ core} = E_{\eta\ clad} \tag{4.16}$$

$$H_{\eta\ core} = H_{\eta\ clad} \tag{4.17}$$

All the field components at the boundary involve both types of Mathieu function. The radial Mathieu functions have arguments

$$C_{en}(\xi_0, \gamma^2_{core}); \ S_{en}(\xi_0, \gamma^2_{core})$$

$$Fek(\xi_0, \gamma^2_{clad}); Gek(\xi_0, \gamma^2_{clad})$$

$Fek_0(\xi, \gamma^2)$ for various γ^2

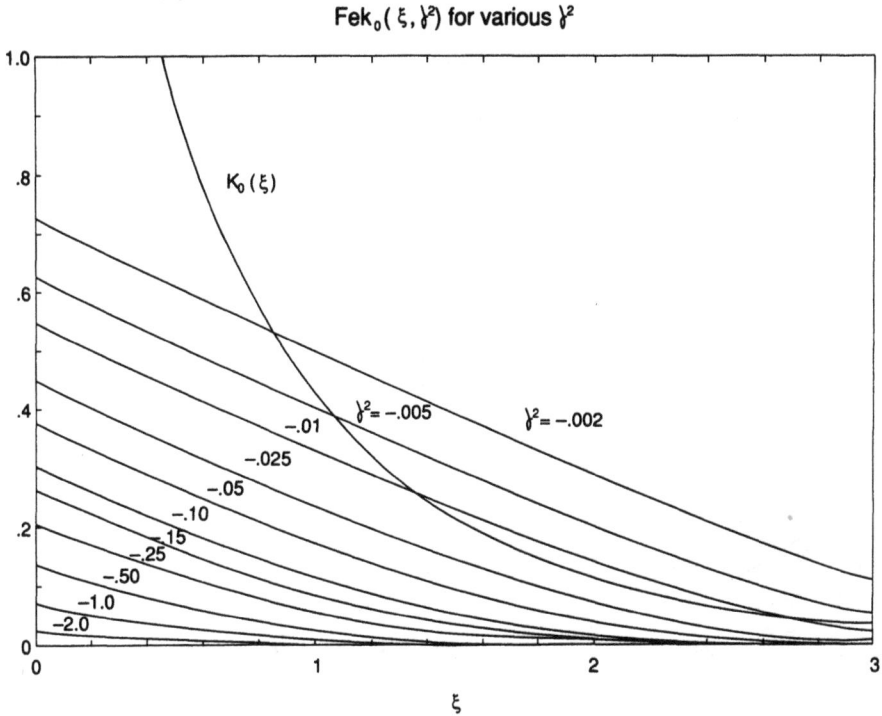

Figure 4.2(e) Mathieu evanescent radial function Fek_0.

The azimuthal functions have arguments

$$c_{en}(\eta, \gamma^2_{core}); \quad s_{en}(\eta, \gamma^2_{core})$$

$$c_{en}(\eta, \gamma^2_{clad}); \quad s_{en}(\eta, \gamma^2_{clad})$$

Therefore, the azimuthal functions (equivalent to $\cos\phi$, $\sin\phi$ in the circular case) have γ^2 and hence ε_1 and ε_2 incorporated into their arguments. This means that it is not possible to match the boundary fields with one set of functions, an infinite order of angular Mathieu functions being needed to describe the fields both in the core and in the cladding.

In the physical sense, all concentric circles have the same shape as do all coaxial infinite slabs, whereas confocal ellipses do not. This alone makes the solution to the elliptical dielectric rod much more complicated than that for the circular rod or infinite slab.

As a demonstration, drop a matchstick broadside onto a water surface and watch the ripples become less elliptical and more circular as they spread.

$$\text{Fek}_1(\xi, \gamma^2)$$

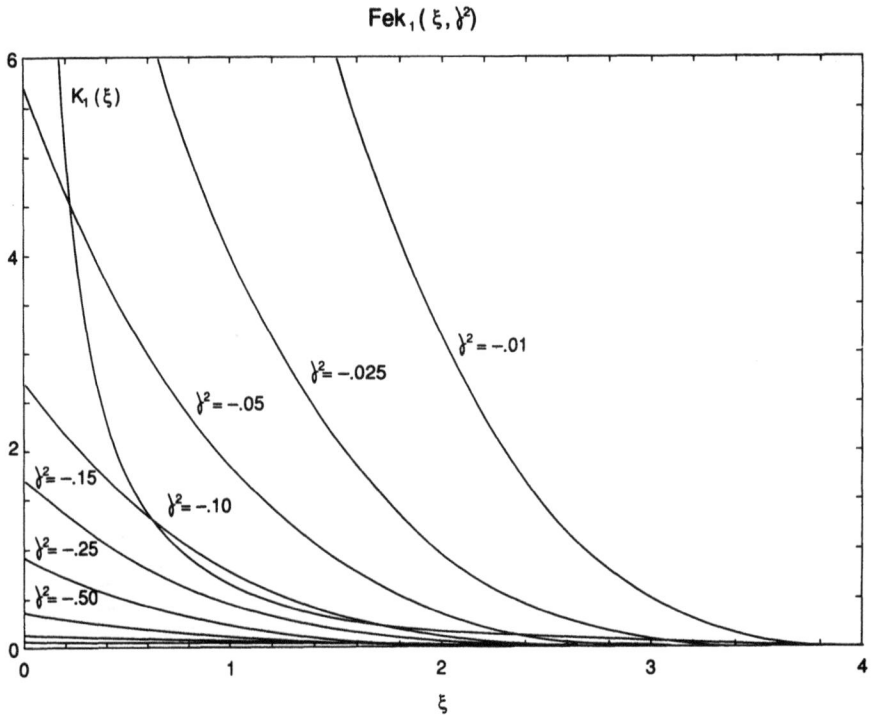

Figure 4.2(f) Mathieu evanescent radial function Fek_1.

Yeh [1] simplifies the problem somewhat by postulating that only one set of Mathieu functions represents the fields in the core and that an infinite series of Mathieu functions represent the fields in the cladding or vice versa. The resultant infinite determinant is solved by a method of successive approximations, a complicated procedure whose description is beyond the scope of this book but which is contained in Yeh's original thesis [5] or as an appendix in Kapany and Burke [6].

4.2 MODES

Another consequence of the azimuthal function being dependent on the core and cladding dielectric constants is that all modes are hybrid HE or EH in the elliptical guide in contrast to the circular guide where the azimuthally symmetric H_{0n} and E_{0n} modes have, respectively, no E_z or H_z components (a fact mentioned as early as 1942 by Shelkunoff [7]).

The fundamental mode in the circular guide, the HE_{11}, splits into two fundamental

$$Gek_1 (\xi, \gamma^2)$$

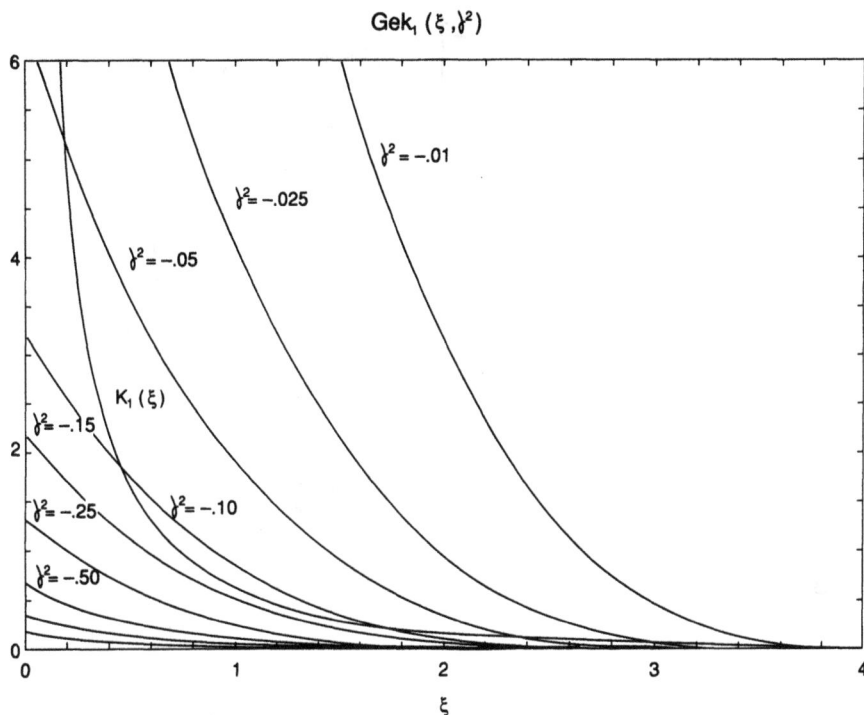

Figure 4.2(g) Mathieu evanescent radial function Gek_1.

modes as the core becomes elliptical. These are the odd HE_{11}, designated $_oHE_{11}$, and even $_eHE_{11}$ where the axial magnetic fields involve the odd and even Mathieu functions, respectively. In common with the circular HE_{11} mode, both $_oHE_{11}$ and $_eHE_{11}$ modes have no cutoff and have transverse electric fields, the $_oHE_{11}$ along the major axis, and the $_eHE_{11}$ along the minor axis of the ellipse. Figure 4.3 shows the field patterns of these fundamental modes. As the ellipse tends to the infinite slab, the $_oHE_{11}$ becomes the H_{10} mode and the $_eHE_{11}$ the E_{10} mode. Figure 4.4 illustrates the transition.

4.3 TRANSCENDENTAL EQUATIONS

Instead of considering the computed exact solution, it is worthwhile to look at a considerable simplification (Yeh [1]) that has only the first four elements of the matrix, equivalent to using one Mathieu function for the core and one for the cladding. This solution applies only to the HE_{1m} modes at ellipticities not exceeding $a/b > 2.5$ and is supposedly limited to a small difference in dielectric constant between core and cladding.

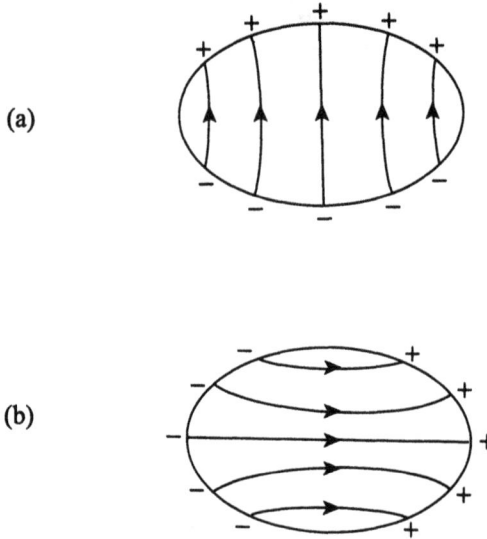

Figure 4.3 Transverse electric field for (a) $e\mathrm{HE}_{11}$ mode, and (b) $o\mathrm{HE}_{11}$ mode.

With these assumptions, the transcendental equations can be written for the odd modes as

$$
\left[\frac{1}{u^2}\frac{S'_{en}(\xi_0,\,\gamma_1^2)}{S_{en}(\xi_0,\,\gamma_1^2)} + \frac{1}{w^2}\frac{Gek'_n(\xi_0,\,-\gamma_2^2)}{Gek_n(\xi_0,\,-\gamma_2^2)}\right]\left[\frac{\varepsilon_1}{u^2}\frac{C'_{en}(\xi_0,\,\gamma_1^2)}{C_{en}(\xi_0,\,\gamma_1^2)} + \frac{\varepsilon_2}{w^2}\frac{Fek'_n(\xi_0,\,-\gamma_2^2)}{Fek_n(\xi_0,\,-\gamma_2^2)}\right]
$$

$$
= n^2\left(\frac{1}{u^2} + \frac{1}{w^2}\right)\left(\frac{\varepsilon_1}{u^2} + \frac{\varepsilon_2}{w^2}\right)
$$

(4.18)

and for the even modes as

$$
\left[\frac{1}{u^2}\frac{C'_{en}(\xi_0,\,\gamma_1^2)}{C_{en}(\xi_0,\,\gamma_1^2)} + \frac{1}{w^2}\frac{Fek'_n(\xi_0,\,-\gamma_2^2)}{Fek_n(\xi_0,\,-\gamma_2^2)}\right]\left[\frac{\varepsilon_1}{u^2}\frac{S'_{en}(\xi_0,\,\gamma_1^2)}{S_{en}(\xi_0,\,\gamma_1^2)} + \frac{\varepsilon_2}{w^2}\frac{Gek'_n(\xi_0,\,-\gamma_2^2)}{Gek_n(\xi_0,\,-\gamma_2^2)}\right]
$$

(4.19)

$$
= n^2\left(\frac{1}{u^2} + \frac{1}{w^2}\right)\left(\frac{\varepsilon_1}{u^2} + \frac{\varepsilon_2}{w^2}\right)
$$

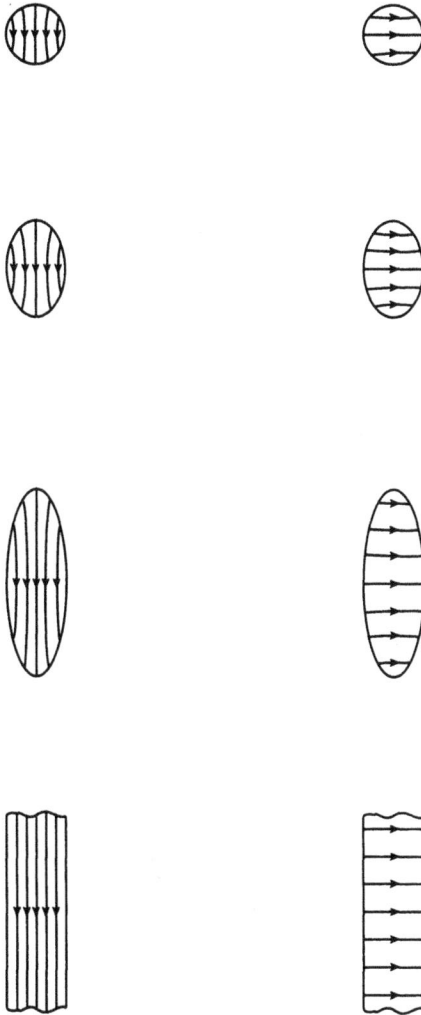

Figure 4.4 Transverse electric field for the fundamental modes of circular, elliptical, and slab waveguide.

with

$$\gamma_1 = \gamma_{core}; \ \gamma_2 = \gamma_{clad} \tag{4.20}$$

and

$$\gamma_1^2 = \frac{q^2}{4}(k_1^2 - \beta^2) = \frac{1}{4}\left(\frac{a^2}{b^2} - 1\right)u^2 \tag{4.21}$$

$$\gamma_2^2 = \frac{q^2}{4}(\beta^2 - k_2^2) = \frac{1}{4}\left(\frac{a^2}{b^2} - 1\right)w^2 \tag{4.22}$$

where

$$k_1 = \sqrt{\varepsilon_1}k_0; \ k_2 = \sqrt{\varepsilon_2}k_0 \tag{4.23}$$

$$u^2 = b^2(k_1^2 - \beta^2) \tag{4.24}$$

$$w^2 = b^2(\beta^2 - k_2^2) \tag{4.25}$$

$$u^2 + w^2 = V_b^2 = k_0^2 b^2(\varepsilon_1 - \varepsilon_2) \tag{4.26}$$

The equations closely resemble those for the circular guide (3.1) and for the slab (2.1, 2.2). As the ellipse becomes a circle

$$\frac{C_{en}'(\xi_0, \gamma_1^2)}{C_{en}(\xi_0, \gamma_1^2)} \text{ and } \frac{S_{en}'(\xi_0, \gamma_1^2)}{S_{en}(\xi_0, \gamma_1^2)}$$

go to

$$\frac{uJ_n'(u)}{J_n(u)}$$

And

$$\frac{Fek_n'(\xi_0, -\gamma_2^2)}{Fek_n(\xi_0, -\gamma_2^2)} \text{ and } \frac{Gek_n'(\xi_0, -\gamma_2^2)}{Gek_n(\xi_0, -\gamma_2^2)}$$

go to

$$\frac{wK_n'(w)}{K_n(w)}$$

As the ellipse becomes a slab

$$C_{en}(\xi_0, \gamma_1^2) \text{ goes to } \cos u$$

$$S_{en}(\xi_0, \gamma_1^2) \text{ goes to } \sin u$$

$$Fek_n(\xi_0, -\gamma_2^2) \text{ goes to } e^{-w}$$

$$Gek_n(\xi_0, \gamma_2^2) \text{ goes to } e^{-w}$$

As with the slab, the propagation constant of the mode having the electric field along the major axis of the ellipse, the $_oHE_{11}$, is greater than that of the $_eHE_{11}$ with the electric field parallel to the minor axis. The $_oHE_{11}$ has the greater "binding geometry" (Yeh [1]), illustrated by Figure 4.5 which shows diagrammatically the distortion of the E field between two parallel metal plates caused by placing a dielectric slab perpendicular to and parallel to the field.

The $_oHE_{11}$ mode therefore has the smaller phase velocity and is more resistant to radiation caused by bending the guide. The modes are sometimes called slow and fast, which are arguably more descriptive terms than the arcane odd and even.

Equation (4.18) and (4.19) may be solved by computer using routines to generate the Mathieu functions [3,4].

A direct comparison can be made between the results obtained by this program and those from an experiment using a microwave resonator (to be described later). Figure 4.6 compares the computed and experimental results for the normalized propagation constants $o\bar{\beta}$ and $e\bar{\beta}$ of the odd and even modes.

There is fairly good agreement between the computed and experimental results to an ellipticity $b/a = 0.4$ even though the dielectric difference $\varepsilon_1 - \varepsilon_2$ is great (very much larger than that in an optical fiber). Unlike the circular guide where the core is circular to within close limits, the shape of the core in the elliptical fiber is not an exact ellipse. The usefulness of the analysis for the fundamental modes is then mainly for an approximate solution together with some physical insight. The argument is reinforced by the fact that

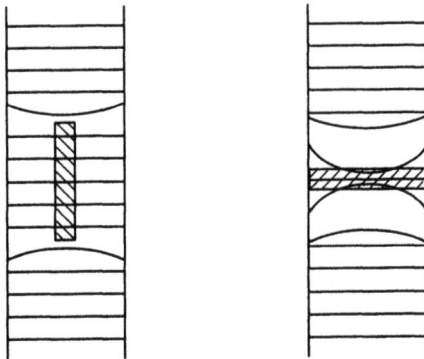

Figure 4.5 Distortion of electric field.

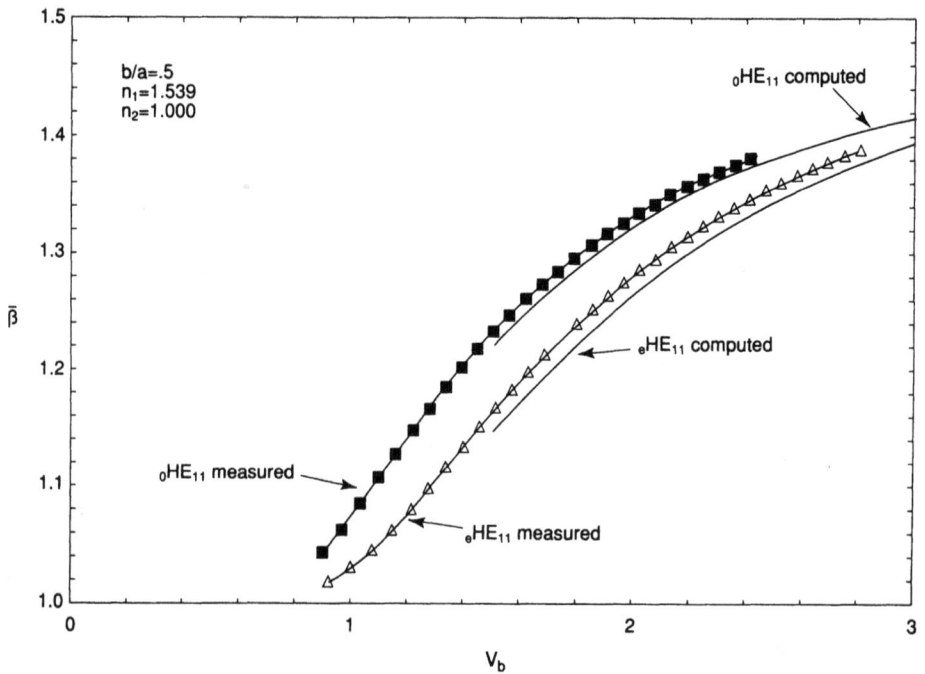

Figure 4.6 Computed and experimental values of $o\bar{\beta}$ and $e\bar{\beta}$.

there is some diffusion of the core-cladding boundary as the fiber is being pulled and that, depending on the method of manufacture, there is probably also a "dip" in the index profile on the fiber axis.

4.4 BIREFRINGENCE

One of the most important characteristics of elliptically cored fiber is its ability to preserve the polarization of a guided wave by the decoupling of the propagation constants of the $_o\text{HE}_{11}$ and $_e\text{HE}_{11}$ modes, otherwise known as the birefringence, and defined as the difference between the normalized propagation constants, $_o\bar{\beta} - _e\bar{\beta} = \Delta\bar{\beta}$.

Defined in this way, the birefringence is the difference in the modal effective indices $_o n_e$ and $_e n_e$. It is a function of V_b and also of the core-cladding index difference Δn. The first analyses of birefringence in elliptical fiber were made in order to estimate the effects of fiber cores that were slightly out of round; for instance, for those fiber communications where differential group delay restricts the system bandwidth and in sensor applications where zero birefringence is necessary (as in current sensors using Faraday rotation). A useful summary of these analyses, given by Adams et al. [8], is as follows.

Ramaswamy et al. [9] approached the problem by adapting Marcatili's calculations for rectangular waveguide [10]. For small eccentricities and small index difference, they have

$$\frac{a\Delta\beta}{e^2\left(1-\left(\frac{n_2}{n_1}\right)^2\right)^{\frac{3}{2}}} \approx \frac{3\pi V^2}{(V+2)^4} \tag{4.27}$$

where:

$$a = \text{core half-width}$$

$$n_1 = \text{core index}$$

$$n_2 = \text{clad index}$$

$$e = \left[1-\left(\frac{b}{a}\right)^2\right]^{\frac{1}{2}}$$

$$V = \frac{2\pi a}{\lambda_0}\left[n_1^2 - n_2^2\right]^{\frac{1}{2}}$$

$$\lambda_0 = \text{free space wavelength}$$

Schlosser [11], using a more sophisticated analysis, has

$$\frac{a\Delta\beta}{e^2\left(1-\left(\frac{n_2}{n_1}\right)^2\right)^{\frac{3}{2}}} \approx \frac{u^4 w^3}{8V^5 J_1^2(u)}\left[\frac{K_0(w)}{K_1(w)}+\frac{1}{w}\right] \tag{4.28}$$

Marcuse [12] has the approximation

$$\frac{a\Delta\beta}{e^2\left(1-\left(\frac{n_2}{n_1}\right)^2\right)^{\frac{3}{2}}} \approx \frac{u^2 w^2}{8V^3} \tag{4.29}$$

Snyder and Young [13] use another approximation to give

$$\frac{a\Delta\beta}{e^2\left(1 - \left(\frac{n_2}{n_1}\right)^2\right)^{\frac{3}{2}}} \approx \frac{u^2 w^2}{8V^3}\left[1 + \frac{uK_0^2(w)J_2(u)}{K_1^2(w)\,J_1\,(u)}\right] \tag{4.30}$$

Adams et al. [8] also use a simplified form of Yeh's analysis and substitute Bessel function approximations for the Mathieu functions to give

$$\frac{a\Delta\beta}{e^2\left(1 - \left(\frac{n_2}{n_1}\right)^2\right)^{\frac{3}{2}}} \approx \frac{u^2 w^2}{8V^5}\left[\left(\frac{J_0(u)}{J_1(u)}\right)^3\left(\frac{(u^2 - w^2)w^2}{u}\right)\right.$$

$$\left. + \left(\frac{J_0(u)}{J_1(u)}\right)^2\left(\frac{w^4 + u^4}{u^2}\right) + \left(\frac{J_0(u)}{J_1(u)}\right)2u\,(4 + w^2) - (8 + w^2 - u^2)\right] \tag{4.31}$$

All these approximations use the normalized birefringence term

$$X = \frac{a\Delta\beta}{e^2\left(1 - \left(\frac{n_2}{n_1}\right)^2\right)^{\frac{3}{2}}} \tag{4.32}$$

which can be arranged in the form

$$\frac{\Delta\bar\beta}{(\Delta n)^2} = \frac{4\frac{b}{a}\left[1 - \left(\frac{b}{a}\right)^2\right]X}{n_1 V_b} \tag{4.33}$$

The birefringence $\Delta\bar\beta$ is then seen to be proportional to $(\Delta n)^2$, which means that for elliptical core fibers a large core-cladding index difference is necessary for the high birefringence associated with maintaining polarization.

The results of these analyses, shown in Figure 4.7, vary widely both in the magnitude of the birefringence and in the value of V where the birefringence is a maximum. Also plotted on Figure 4.7 are values produced by the computer solution mentioned earlier in the chapter and modeling a weakly guiding fiber $n_1 = 1.5$, $\Delta n = .01$ with small eccentricity $b/a = 0.9$, $e = .435$. Best agreement is with the results of Adams and others [8].

Figure 4.8 from Dyott et al. [14] shows $\frac{\Delta\bar\beta}{(\Delta n)^2}$ against V_b for ellipticities varying

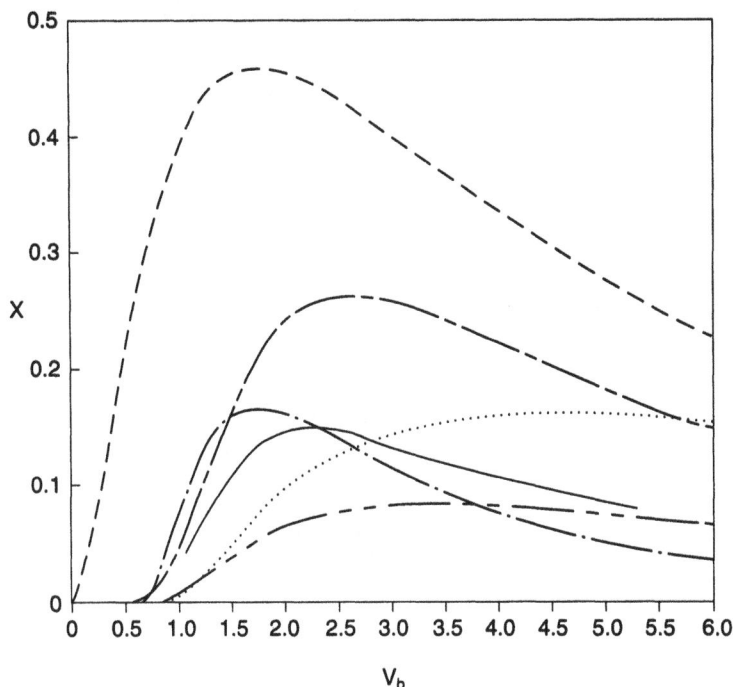

Figure 4.7 Comparison of several methods for calculating birefringence of weakly guiding fibers with small ellipticity: — · — · —, Adams et al. [8]; — — — —, Ramaswamy et al. [9]; — · — · · —, Schlosser [11]; — · · — · · —, Marcuse [12]; · · · · · ·, Snyder and Young [13]; ——————, computed Mathieu function solution.

from circle to slab. These values have been calculated using the four-term matrix approximation described above.

The broken line is the locus of the higher mode cutoff, the fundamental mode region being to the left of the line with the lower values of V_b.

The group velocities, expressed as the group index $n_g = \dfrac{c}{v_g}$, of the fundamental modes are shown in Figure 4.9 and the difference in group index $\Delta n_g = {}_on_g - {}_en_g$ in Figure 4.10 from [14]. As with the slab at a particular value of V_b, depending here on the ellipticity, $\Delta n_g = 0$ and for higher values of V_b, ${}_on_g < {}_en_g$ so that the group velocity of the slow ${}_o\text{HE}_{11}$ mode becomes greater than that of the fast ${}_e\text{HE}_{11}$ mode. The phase velocities however remain ${}_ov_p < {}_ev_p$.

The point of zero group delay difference, which could have been so important for communication fibers with slightly out-of-round cores, is unfortunately always in the multimode region above higher mode cutoff. Figure 4.11 shows V_b for zero difference in group delay from $0.2 < b/a \leq 1.0$. Also plotted is the cutoff for higher order modes.

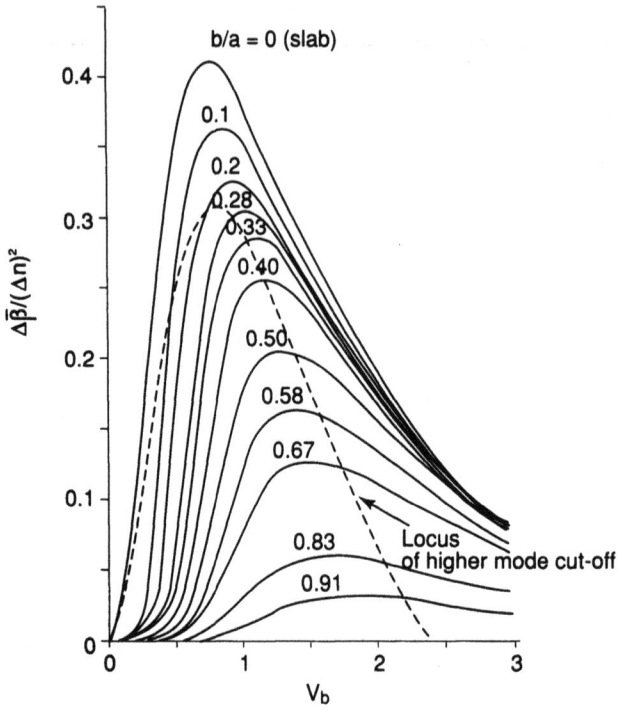

Figure 4.8 Normalized birefringence of elliptical dielectric waveguide.

4.5 POWER DISTRIBUTION

The easiest way of computing the power distribution in the elliptical guide is again to use

$$\frac{P_{core}}{P_{total}} = \frac{\dfrac{\bar{\beta}}{v_g} - n_2^2}{n_1^2 - n_2^2} \qquad (4.34)$$

Figure 4.12 shows $\eta = \dfrac{P_{core}}{P_{total}}$ and Figure 4.13 the differential power distribution between modes normalized with respect to Δn as a function of ellipticity. The $_o\mathrm{HE}_{11}$ mode with the better "binding geometry" has the greater fraction of power in the core.

4.6 A PERTURBATION APPROACH

As an alternative to the direct but intricate Mathieu function analysis, Kumar and Varshney [15] have used the first order perturbation of a rectangular dielectric waveguide,

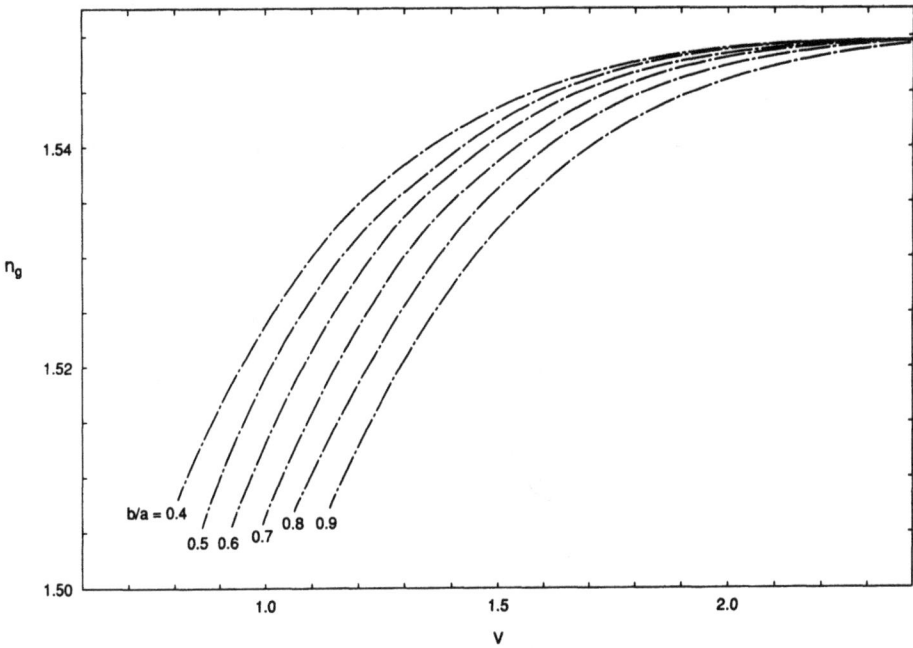

Figure 4.9 Group index for elliptical fiber.

shown in Figure 4.14. The rectangular cross section has the same area and ratio of major/minor axes as the ellipse, so that if the semimajor and semiminor axes of the ellipse and rectangle are a_e, b_e, a_r, b_r, then, equating areas

$$\pi a_e b_e = 4 a_r b_r \tag{4.35}$$

Equating aspect ratios

$$\frac{a_e}{a_r} = \frac{b_e}{b_r} \tag{4.36}$$

$$\frac{a_e}{a_r} = \frac{b_e}{b_r} = \frac{1}{2} (\pi)^{\frac{1}{2}} \tag{4.37}$$

The refractive index of the innermost area bounded by ellipse and rectangle is the core index n_1 and outside that area n_2, the cladding index.

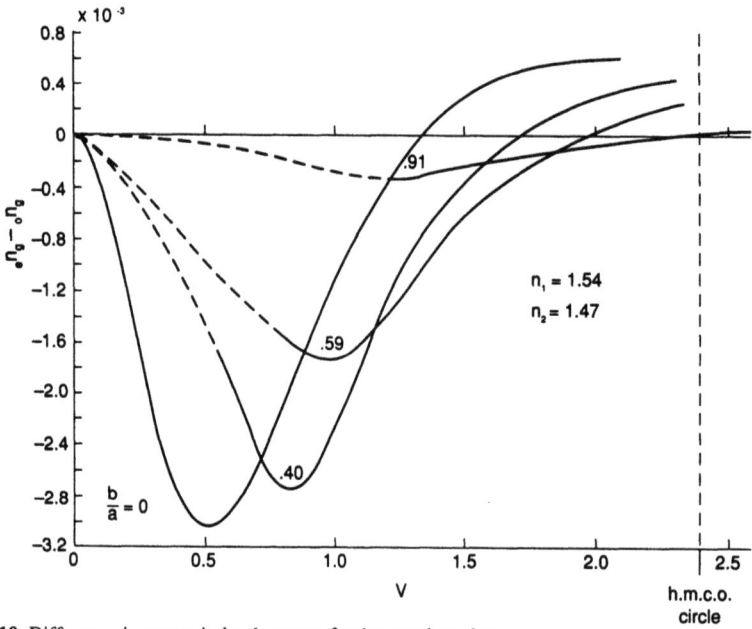

Figure 4.10 Difference in group index between fundamental modes.

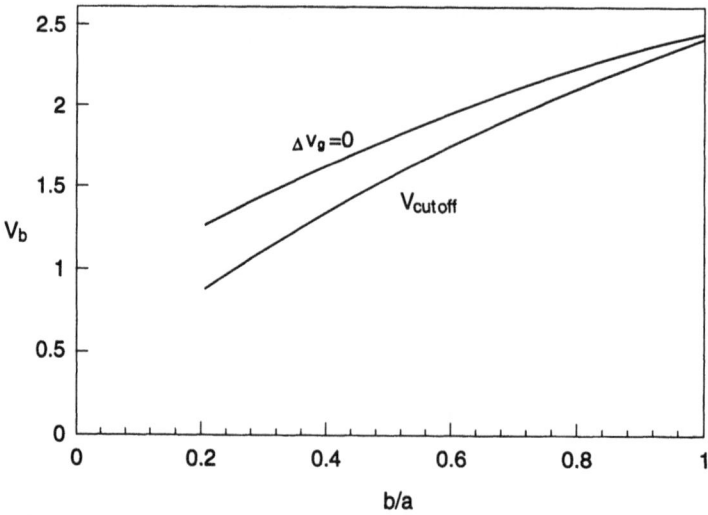

Figure 4.11 Condition for zero difference in group delay.

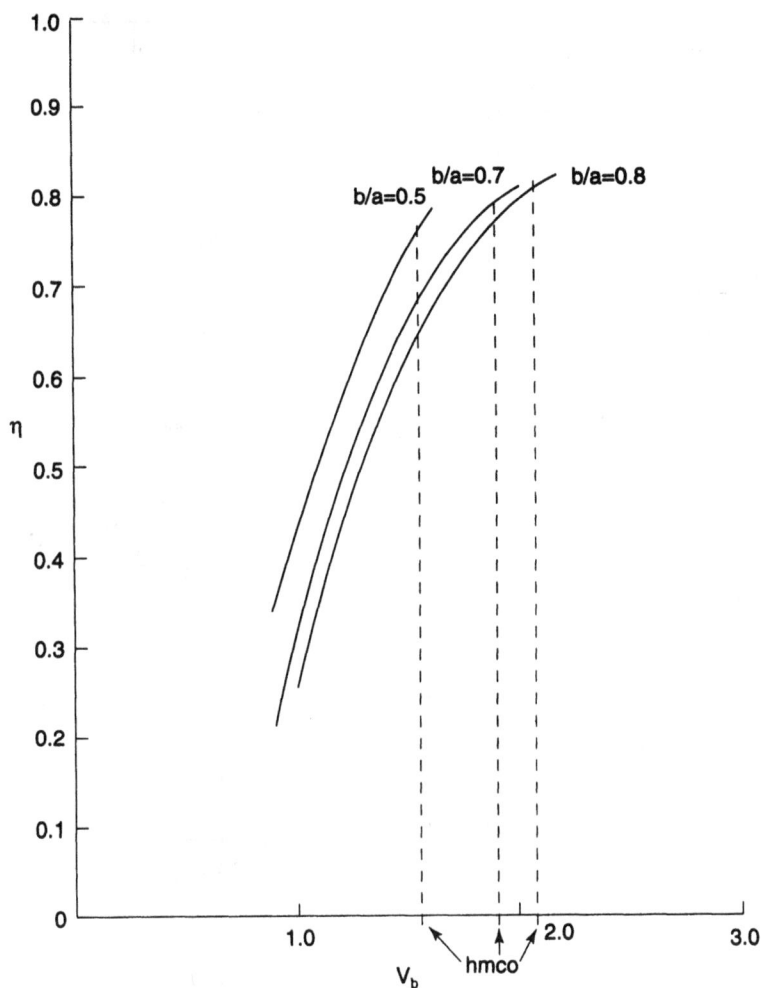

Figure 4.12 Power distribution for small Δn.

Matching fields at the boundary produces transcendental equations with no more complicated a function than Tan^{-1}.

The propagation constants of the elliptical guide are found by perturbing the index by δ_n in the four regions marked in the figure such that

$$\delta_n^2 = n_1^2 - n_2^2 \text{ in regions 1, 3, and 4} \qquad (4.38)$$

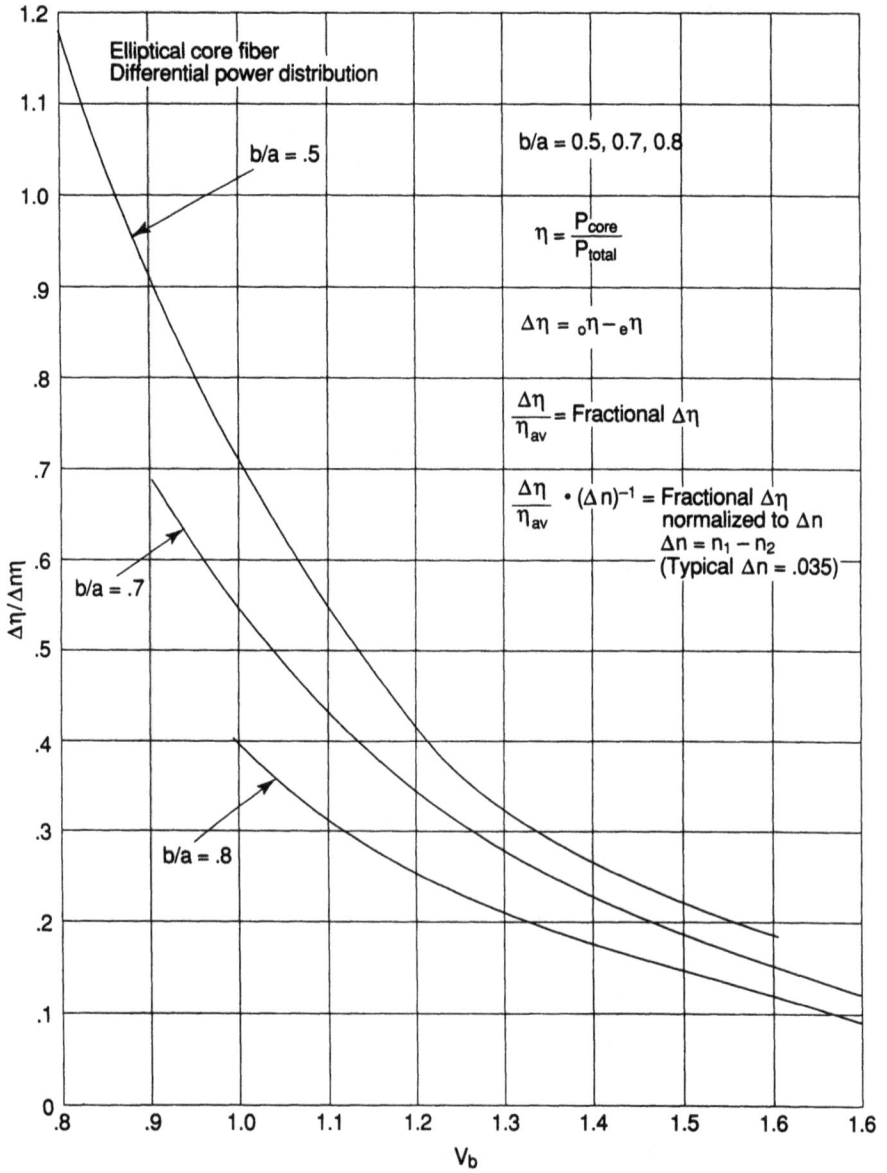

Figure 4.13 Difference in power distribution between fundamental modes.

$$\delta_n^2 = n_2^2 - n_1^2 \text{ in region 2} \tag{4.39}$$

Some complicated but straightforward integration and manipulation eventually produce the propagation constants of the fundamental and higher order modes. Figure 4.15 shows the propagation parameter $\dfrac{\bar{\beta}^2 - n_2^2}{n_1^2 - n_2^2}$ of their $E_{11}y$ mode which is polarized in the y direction

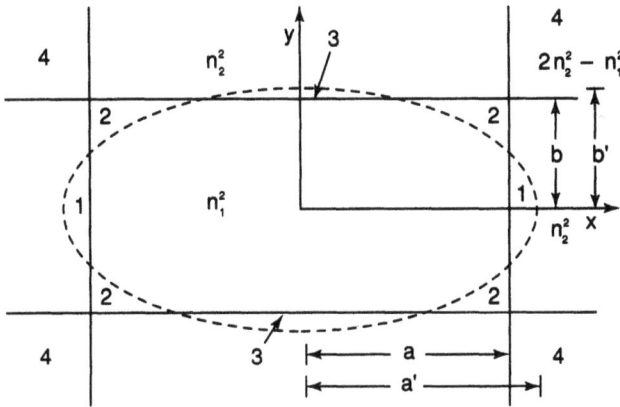

Figure 4.14 Geometry and regional index for perturbation method.

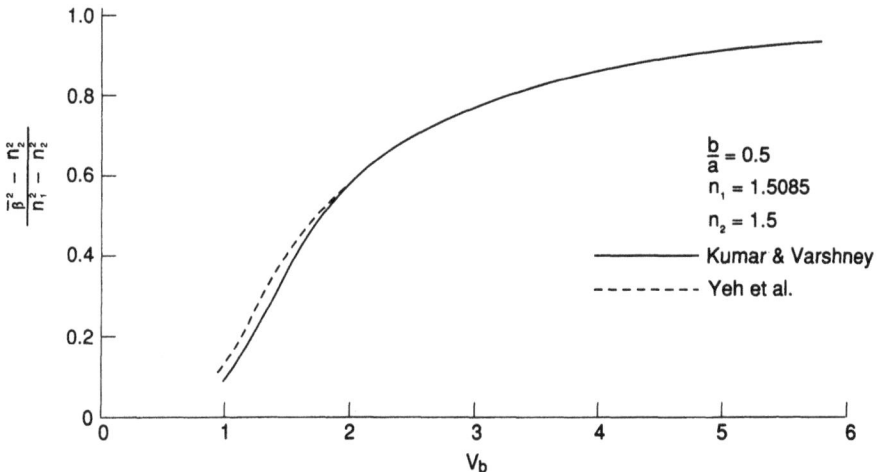

Figure 4.15 Propagation parameter $\dfrac{\bar{\beta}^2 - n_2^2}{n_1^2 - n_2^2}$ for the E_{11y} mode from Kumar and Varshney.

and corresponds to the $_eHE_{11}$ mode. Also plotted in the same figure are results by Yeh et al. [16] using a finite element technique. The agreement between the two methods is fairly good, especially at the lower values of V_b. However, the index difference $\Delta n = 0.0085$ is lower than that of a practical polarization holding fiber by a factor of about 3.5 and inaccuracies increase with Δn.

Figure 4.16 shows $\dfrac{\Delta\bar\beta}{(\Delta n)^2}$ from Kumar and Varshney [15] compared with values from the Mathieu function analysis [14]. The latter are themselves compared in Figure 4.17 with experimental values from the microwave resonator. There is good agreement in spite of the large index difference of the microwave model, $\Delta n = 0.516$.

4.7 FAR-FIELD RADIATION PATTERN

The far-field intensity pattern has been derived using a gaussian function approximation, by Varshney et al. [17]. At the time of writing there does not seem to be any work

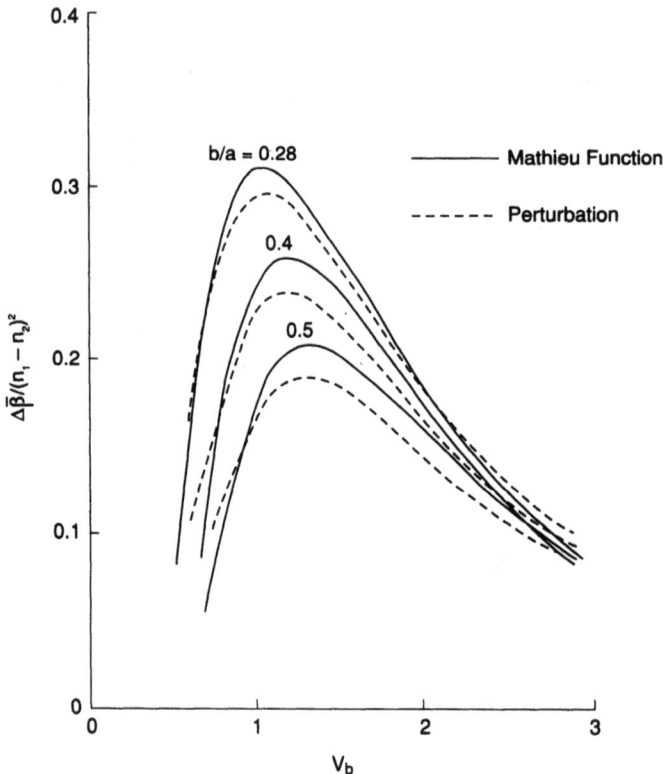

Figure 4.16 $\Delta\bar\beta/(\Delta n)^2$—Comparison of methods.

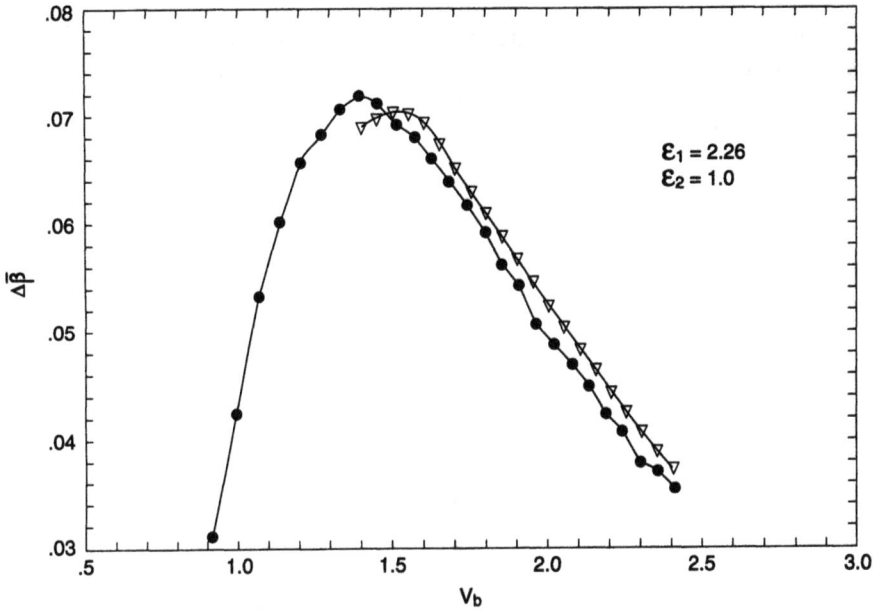

Figure 4.17 Normalized birefringency for large index difference—comparison between Mathieu function analysis and microwave experiment. ∇∇∇, Computed from Mathieu function analysis; •••, measured using microwave resonator.

published in the open literature on the theory of the radiation from elliptical fibers using the Mathieu function analysis. However, a thesis by Boucouvalas [18] calculates the far-field pattern by applying the Fourier transform to the waveguide fields. He arrives at the integral

$$F_x(\alpha,\beta) =$$

$$\frac{q^2}{2\lambda^2} \int_0^\infty \int_0^{2\pi} E_{\alpha x}(\xi, \eta) \exp [jk \, (\alpha q \cosh \xi \cos \eta + \beta q \sinh \xi \sin \eta)]$$

$$(\cosh 2\xi - \cos 2\eta)\partial\xi\partial\eta$$

(4.40)

with

$$\alpha = \sin \theta \cos \phi \tag{4.41}$$

$$\beta = \sin \theta \sin \phi \tag{4.42}$$

in polar coordinates. $E_{\alpha x}$ is the x component of the aperture field at the end of the fiber.

He justly remarks that "This integral is very difficult to evaluate analytically, hence the results were obtained numerically using a fast Fourier subroutine." Figure 4.18 shows his computed far-field radiation pattern along the major and minor axes for the $_o\text{HE}_{11}$ mode with $b/a = 0.5$, $n_1 = 1.54$, and $n_2 = 1.47$ for values of $V_b = 1.6$ and 2.3.

(a)

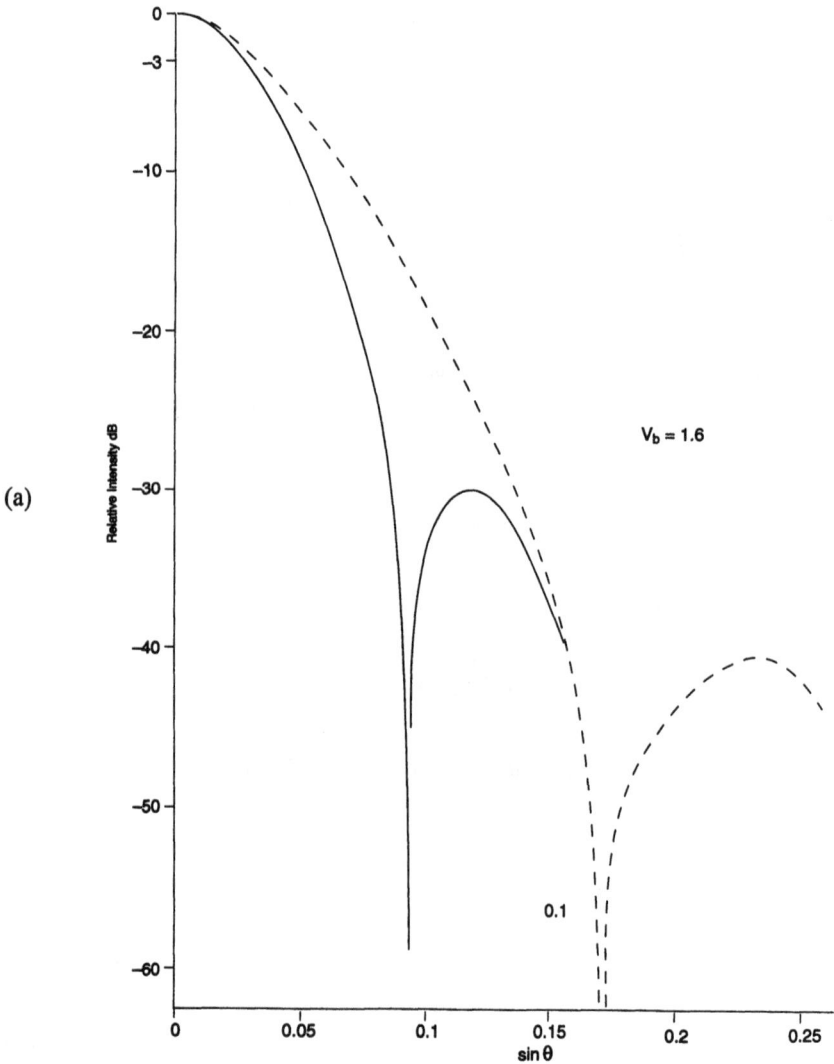

Figure 4.18 Computed far-field radiation pattern of the $_o\text{HE}_{11}$ mode with $n_1 = 1.54$, $n_2 = 1.47$: (a) for $V = 1.6$; (b) for $V = 2.3$;———, Major axis; - - - - -, minor axis.

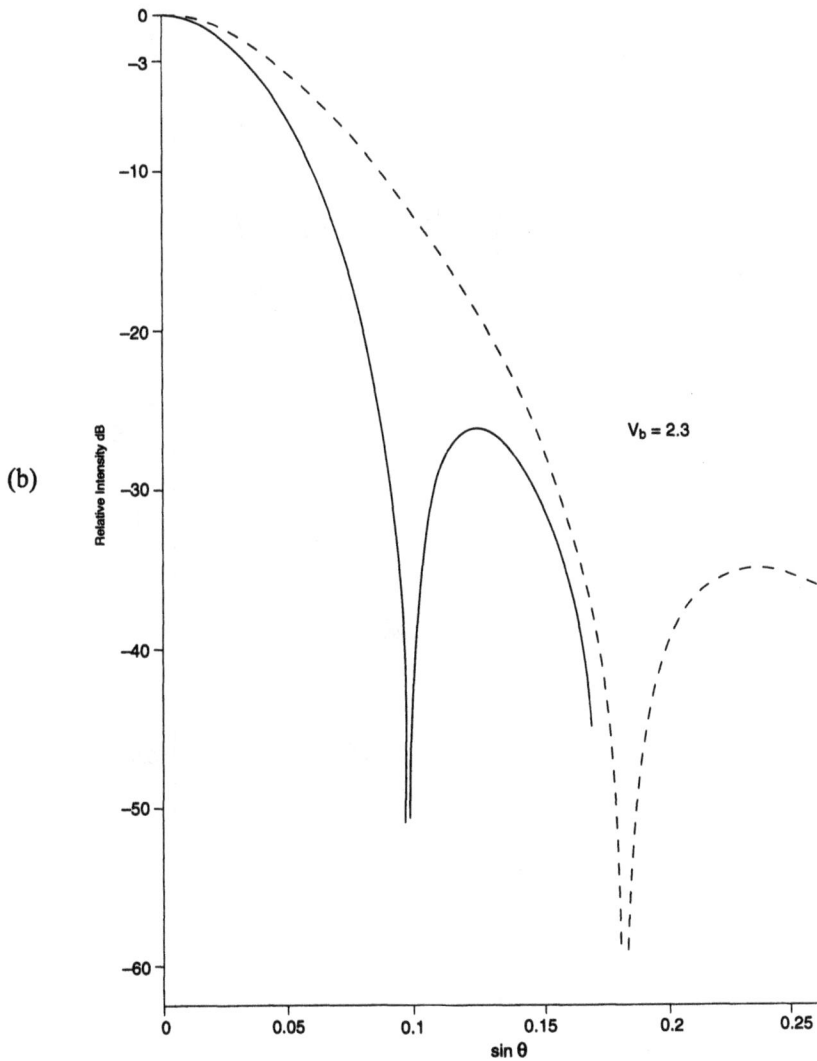

Figure 4.18 (continued)

Figure 4.19 shows Boucouvalas' comparative plot for $b/a = 0.5$ of the relationship between half-power angle Θ_h of the main lobe and the angle to the first minimum Θ_x for the dielectric slab, circular core, and elliptical core. Compare this with Figure 4 from [4], Chapter 3.

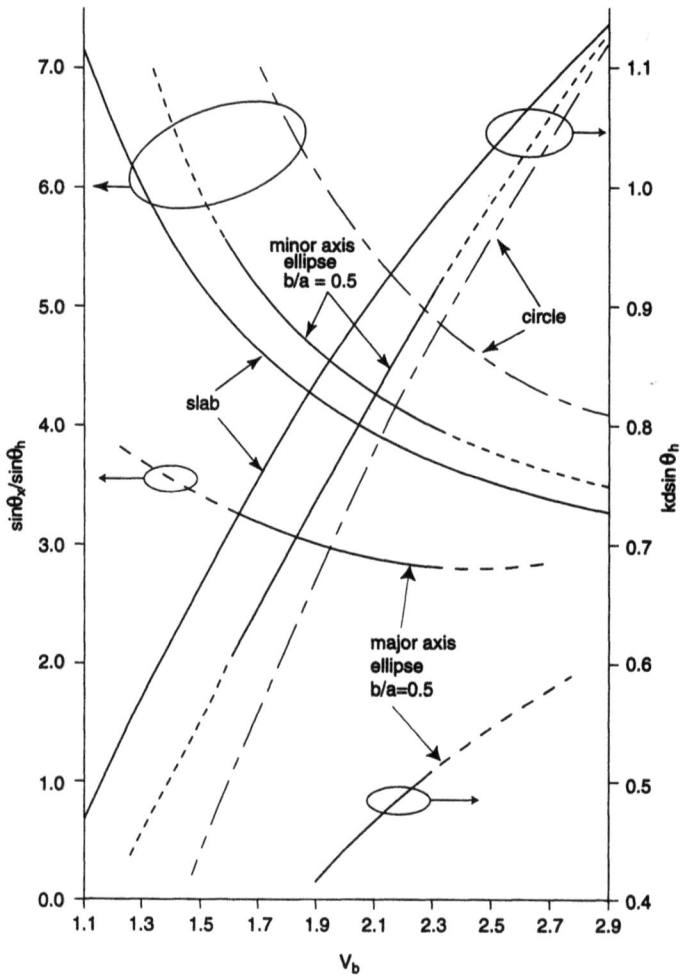

Figure 4.19 Relationships between half-power angle θh of main lobe and angle of first minimum θ for slab, circle, and ellipse, $b/a = 0.5$. (From [18].)

REFERENCES

[1] Yeh, C. "Elliptical dielectric waveguides," *Journal Applied Physics,* Vol. 33, 1962, pp. 3235–3243.

[2] McLachlan, N. W. *Theory and Application of Mathieu Functions,* Oxford University Press: London, 1947.

[3] Rengarajan, S. R., and J. E. Lewis. "Mathieu functions of integral order and real arguments," *IEEE Trans. Microwave Theory and Techniques,* Vol. MTT-28, 1980, pp. 276–277.

[4] Toyama, N., and K. Shogen. "Computation of the value of the even and odd Mathieu Functions of order N for a given parameter S and an argument X," *IEEE Trans. on Antennas and Propagation,* Vol. AP-32, 1984, pp. 537–539.

[5] Yeh, C. "Electromagnetic surface-wave propagation along a dielectric cylinder of elliptical cross section," *Technical Report No. 27,* Antenna Laboratory, California Institute of Technology, Pasadena, California, 1962.

[6] Kapany, H. S., and J. J. Burke. *Optical Waveguides,* Academic Press: New York, 1972.

[7] Shelkunoff, S. A. *Electromagnetic Waves,* Van Nostrand: New York, 1943, p. 425.

[8] Adams, M. J., D. N. Payne, and C. M. Ragdale. "Bi-refringence in optical fibers with elliptical cross section," *Electronics Letters,* Vol. 15, 1979, pp. 298–299.

[9] Ramaswamy, V., W. G. French, and R. D. Standley. "Polarization characteristics of non-circular single mode fibers," *Applied Optics,* Vol. 17, 1978, pp. 3014–3017.

[10] Marcatili, E.A.J. "Dielectric optical waveguides and directional coupler for integrated optics," *Bell System Technical Journal,* Vol. 48, 1969, pp. 2071–2102.

[11] Schlosser, W. O. "Delay distortion in weakly guiding optical fibers due to elliptic deformation at the boundary," *Bell System Technical Journal,* Vol. 51, 1972, pp. 487–492.

[12] Marcuse, D. *Theory of Dielectric Optical Waveguides,* Academic Press: New York, 1974, p. 158.

[13] Snyder, A. W., and W. R. Young. "Modes of optical waveguides," *Journal Optical Society of America,* Vol. 68, 1978, pp. 297–309.

[14] Dyott, R. B., J. R. Cozens, and D. G. Morris. "Preservation of polarization in optical fiber waveguides with elliptical cores," *Electronics Letters,* Vol. 15, 1979, pp. 380–382.

[15] Kumar, A., and R. K. Varshney. "Propagation characteristics of highly elliptical core optical waveguides: a perturbation approach," *Optical and Quantum Electronics,* Vol. 16, 1984, pp. 349–354.

[16] Yeh, C., K. Ha, B. Dong, and W. P. Brown. "Single mode optical waveguides." *Applied Optics,* Vol. 18, 1979, p. 1490.

[17] Varshney, R. K., R. Srivastava, and R. V. Ramaswamy. "Characterization of highly elliptical submicron core polarization preserving fibers: theory and experiment," *Applied Optics,* Vol. 27, 1988, pp. 3114–3120.

[18] Boucouvalas, A. C. "Use of far field radiation pattern to characterize single mode fibers," Thesis submitted for the degree of Master of Science Imperial College, London, Dept. of Electrical Engineering, 1979.

Chapter 5

Higher Order Modes in
Elliptical Dielectric Waveguides

5.1 HIGHER ORDER MODES IN ELLIPTICAL DIELECTRIC WAVEGUIDES

The number of modes in the dielectric slab is limited by the number of possible variations in field in only one direction because there can be no variations in the transverse y direction, which extends to infinity. The condition for modal cutoff is that the periodic function that describes the field inside the core, with V as the argument, is equal to zero; that is,

$$\cos(V) = 0$$

$$\sin(V) = 0$$

$$V = \frac{2\pi}{\lambda_0} [\varepsilon_1 - \varepsilon_2]^{\frac{1}{2}}$$

There are many more modes possible in the circular dielectric guide because there are field variations in both radial and azimuthal directions. Again, the modal cutoff is given by setting the periodic function of V, equal to zero.

$$J_n(V) = 0$$

As soon as the circular guide is made elliptical, the number of possible modes nearly doubles. Modes that are degenerate in the circular guide become nondegenerate in the ellipse because the circular symmetry is broken; for instance, as the fundamental HE_{11} in circular guide splits into the $_oHE_{11}$ and $_eHE_{11}$. The same applies to most higher order modes. Figure 5.1 shows the splitting of the circular HE_{21} mode into $_oHE_{21}$ and $_eHE_{21}$. The H_{01} and E_{01} modes in circular guide become hybrid $_eHE_{01}$ and $_oEH_{01}$ as they lose their azimuthal symmetry. Thus, all modes in elliptical dielectric guide are hybrid.

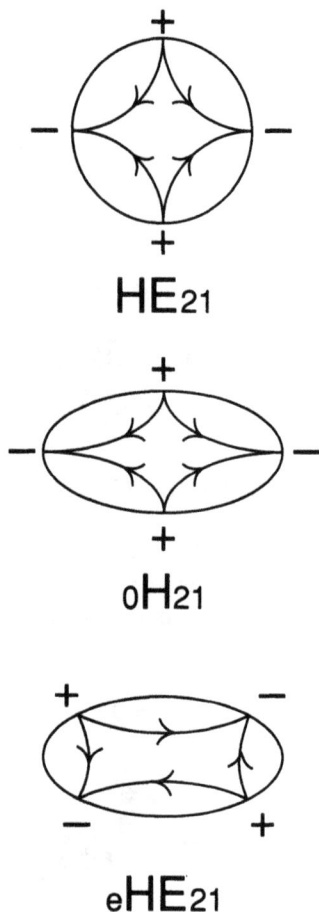

Figure 5.1 HE_{21} mode patterns of transverse electric field for circular and elliptical guide.

Intuitively, it might seem that (as with the slab and the circle) the cutoff conditions for the elliptical guide would be given by putting the periodic functions (that describe the field in the core) equal to zero; that is,

$$S_e(V) = 0$$

$$C_e(V) = 0$$

Unfortunately, the solution is more complicated. The problem, again, is that the argument, γ^2, of the Mathieu functions that describe the azimuthal field, $s_e(\eta, \gamma^2)$, $c_e(\eta, \gamma^2)$, depends on index and is therefore different for core and cladding. Again, an infinite matrix of

Mathieu functions is needed to calculate higher mode cutoff. There is an additional problem in that near cutoff the fields extend far into the cladding, so truncating the matrix to a manageable number of terms can produce inaccurate values.

It is useful, at this stage, to sidestep the problem by considering the metal-walled elliptical waveguide, which has no field beyond the core boundary and where, consequently, the change of γ^2 with index is irrelevant.

5.2 ELLIPTICAL WAVEGUIDES WITH CONDUCTING WALLS

The first published study of the elliptical metal waveguide was by L. J. Chu in 1938 [1]. He derived the field equations and cutoff conditions for several modes, although he missed the even and odd $_eH_{21}$ and $_oH_{21}$ modes. Due to a slip in the computation, the field pattern of the E_{01} mode was plotted incorrectly; the correction was made many years later in 1990 by Goldberg et al. [2] who also corrected discrepancies in cutoff calculations at large ellipticities.

In 1964, Piefke [3] derived new asymptotic formulas for the Mathieu functions and extended the number of classified modes, although he too missed the $_eH_{21}$ and $_oH_{21}$ modes. In 1970, Kretzschmar [4] completed the classification of the first nineteen modes including useful polynomial expressions for the modal cutoffs.

The boundary conditions for the metal-wall waveguide are much simpler than those for the dielectric guide, the sole requirement being that the tangential electric field is zero. This gives, for the cutoffs

$$C_{en}(\xi_0, \gamma^2) = 0$$

E modes $\qquad\qquad\qquad\qquad\qquad\qquad\qquad\qquad$ (5.1)

$$S_{en}(\xi_0, \gamma^2) = 0$$

$$C'_{en}(\xi_0, \gamma^2) = 0$$

H modes $\qquad\qquad\qquad\qquad\qquad\qquad\qquad\qquad$ (5.2)

$$S'_{en}(\xi_0, \gamma^2) = 0$$

(using the nomenclature of Chapter 4.)

Kretzschmar (Figure 6 of [4]) plots λ_c/a against eccentricity e, where C is the cutoff wavelength and a is the semimajor axis. A plot of V_c against b/a for the first 8 modes, calculated from Kretzschmar's polynomials, is shown in Figure 5.2.

The first two modes, $_eH_{11}$ and $_oH_{11}$, correspond to the $_eHE_{11}$ and $_oHE_{11}$ modes in the dielectric guide, which have no cutoff. They both originate from the H_{11} in circular guide at $b/a = 1$ and end up as the E_{10} and H_{10}, respectively, in the parallel plate guide, which is the metal equivalent of the dielectric slab, at $b/a = 0$.

The next higher mode is the E_{01}, which follows the line $C_{e0}(V) = 0$ from $J_0(V) = 0$ at $b/a = 1$ to $\cos(V) = 0$ at $b/a = 0$.

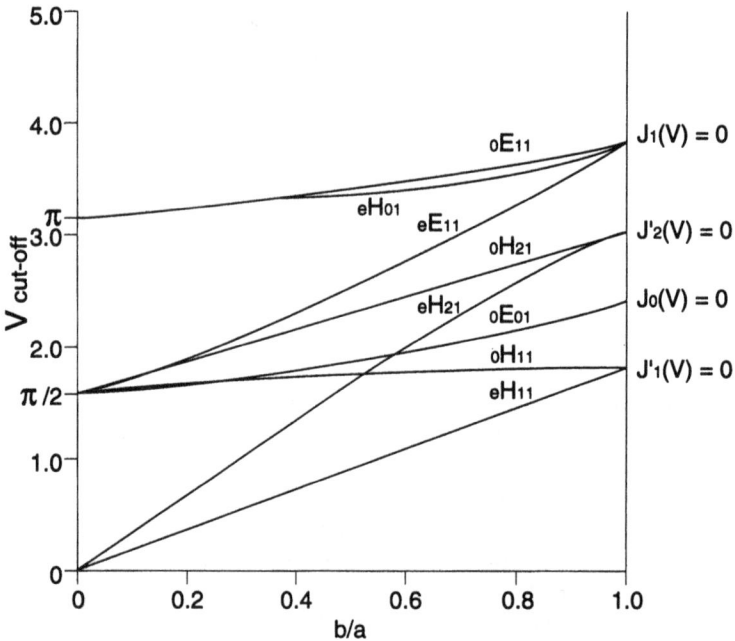

Figure 5.2 Cutoff in elliptical metal waveguides.

The H_{21} (degenerate at $b/a = 1$ with a cutoff at $J_2(V) = 0$, $V = 3.05$) splits into the $_eH_{21}$ and $_oH_{21}$ with cutoffs at $b/a = 0$ of $V = 0$ and $V = \pi/2$, respectively.

Similarly, the E_{11} (circular cutoff, $J_1(V) = 0$, $V = 3.84$) splits, with the $_eE_{11}$ going to $V = \pi/2$ and the $_oE_{11}$ to $V = \pi$ at $b/a = 0$. The H_{01} is also cut off at $J_1(V) = 0$, $V = 3.84$ in circular guide, and also going to $V = \pi$ at $b/a = 0$.

As pointed out by Chu [1], modes with circular symmetry in a circular metal guide, (i.e., H_{0m} E_{0m}) do not split into even and odd in elliptical guide as do all other modes.

5.3 CUTOFF IN ELLIPTICAL DIELECTRIC WAVEGUIDE

In contrast with the metal waveguide, the published work on higher order mode cutoff in elliptical dielectric rods where the dielectric/refractive index difference between core and cladding is large is confined to a few values at small ellipticities.

There are, however, some results obtained experimentally using a microwave resonator formed from an elliptical dielectric rod, terminated at each end with a metal plate. Figure 5.3 shows a sketch of the arrangement. The resonator is weakly coupled to a swept frequency generator at one end and to a detector connected to an oscilloscope at the other. Resonances appear as spikes on the oscilloscope display, whose frequency is measured by

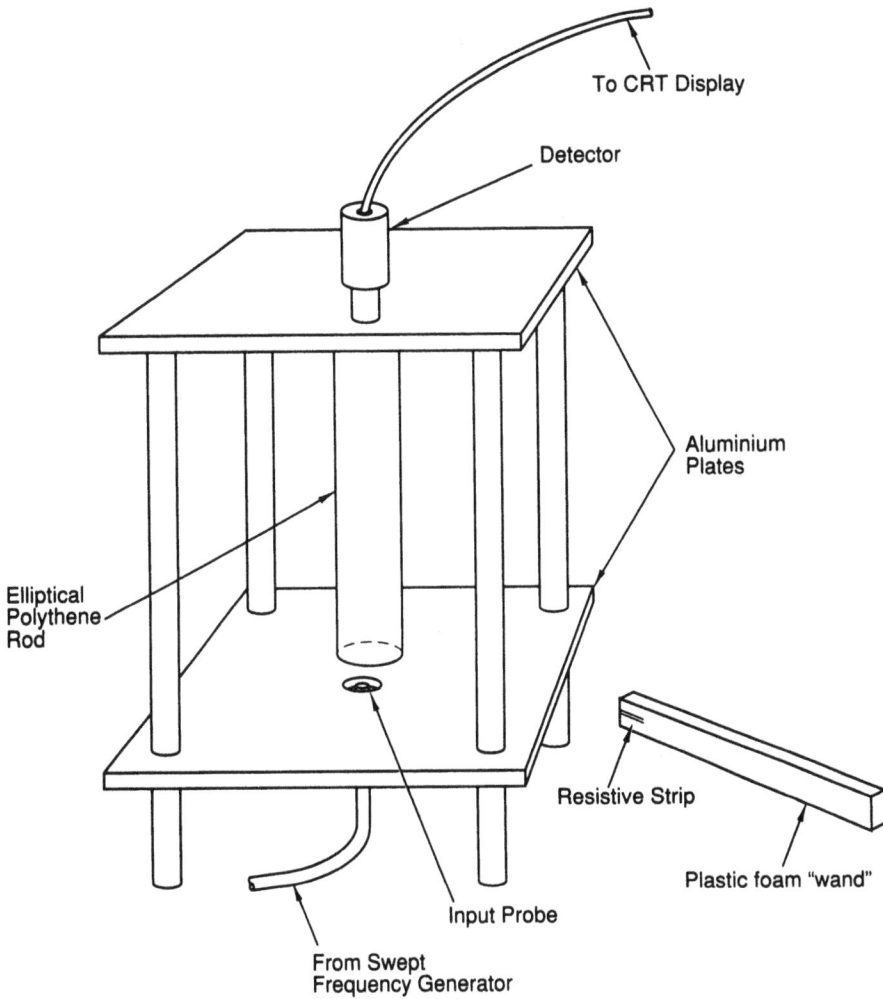

Figure 5.3 Microwave dielectric waveguide resonator.

markers. The number of half-guide wavelengths $\lambda_g/2$ along the rod is found by moving a strip of absorbent material along the rod and counting the number N of extinctions of the displayed spikes as the strip passes through the maxima of the electric field of the standing wave pattern (Dyott [5]).

The resonant frequency, rod dimensions and dielectric constant give

$$V_b = \frac{2\pi b}{\lambda_0} [\varepsilon_1 - \varepsilon_2]^{\frac{1}{2}} \tag{5.3}$$

$$\beta = \frac{2\pi}{\lambda_g} \tag{5.4}$$

$$N\frac{\lambda_g}{2} = L = \text{length of rod} \tag{5.5}$$

The dielectric constant ε_1 is found from the first higher order mode cutoff (at $V = 2.4048$) in a piece of the same circular rod from which all elliptical rods are machined. This ensures that the rod dielectric constants are identical and accurately known.

Because resonant frequencies and rod dimensions can be measured precisely, plots of normalized propagation constant $\overline{\beta} = \beta\frac{\lambda_0}{2\pi}$ against V_b turn out to be surprisingly accurate. Figure 5.4, for instance, shows $\overline{\beta}$ versus V_b for a circular rod, $\varepsilon_1 = 2.3$ in free space $\varepsilon_2 = 1.0$. Plotted for comparison are values calculated from the exact solution of the circular dielectric rod (3.1). Figure 5.5 shows $\overline{\beta}$ versus V_b for an elliptical rod with $b/a = 0.2$ with

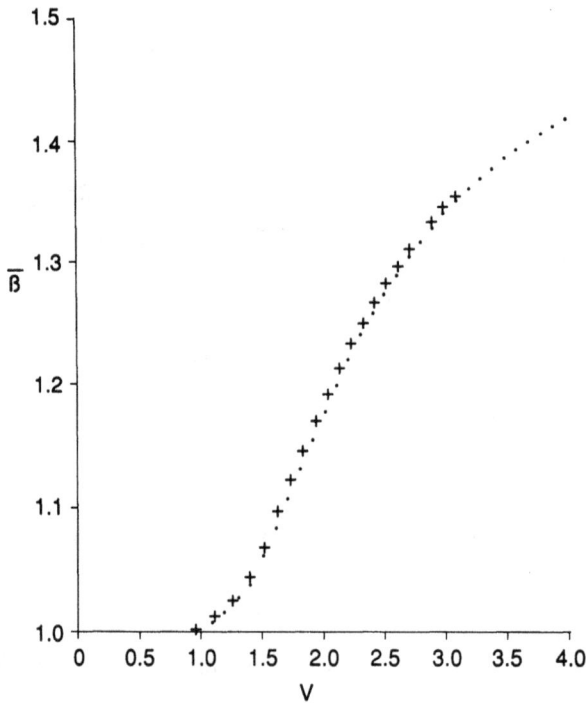

Figure 5.4 Normalized propagation constants of fundamental HE_{11} mode in circular dielectric waveguide for $\varepsilon_1 = 2.30$, $\varepsilon_2 = 1.0$. +, Experimental points; ·, computed points.

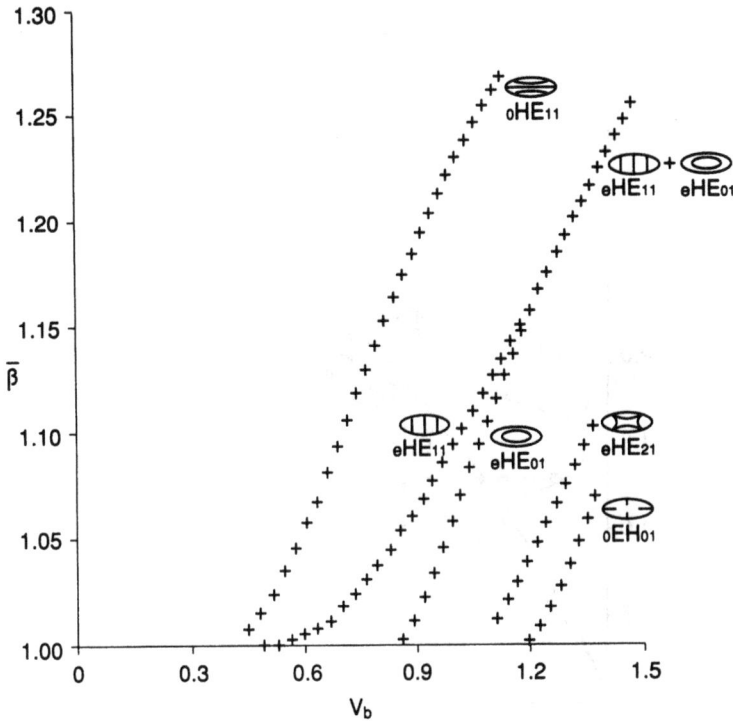

Figure 5.5 Normalized propagation constants of first few modes in elliptical dielectric waveguide. $\varepsilon_1 = 2.3$; $\varepsilon_2 = 1.0$; $b/a = 0.2$. Experimental points from microwave resonator.

the fundamental and higher order mode lines clearly delineated. An idea of the distribution of the electric fields may be found by moving a plane strip of absorbent material azimuthally around the rod and finding the position of field maxima from the extinction of the resonances. Figure 5.6 is a diagram of V_b at cutoff at different ellipticities b/a for the first five higher order modes.

It is interesting to compare this diagram with that of the metal elliptical waveguide. In order to make the comparison clearer, the latter has been redrawn as Figure 5.7, leaving out the fundamental $_oH_{11}$ and $_eH_{11}$ modes since the equivalent dielectric modes ($_oHE_{11}$ and $_eHE_{11}$) have no cutoff.

Comparing the cutoff diagrams, the H_{01} cutoff line in the dielectric guide (Figure 5.6) starts from $V = 2.405$, $J_0(V) = 0$, at $b/a = 1$ and descends to $V = 0$, $\sin(V) = 0$ at $b/a = 0$, compared with the metal guide (Figure 5.7) where it starts from $V = 3.83$, $J_1(V) = 0$ descending to $V = \pi$, $\sin(V) = 0$. This change is caused by the elimination of the extra zero of tangential electric field at the metal boundary. The E_{01} (Figure 5.6) retains its circular cutoff at $V = 2.405$. As b/a decreases, both cutoffs are effectively identical until a split can

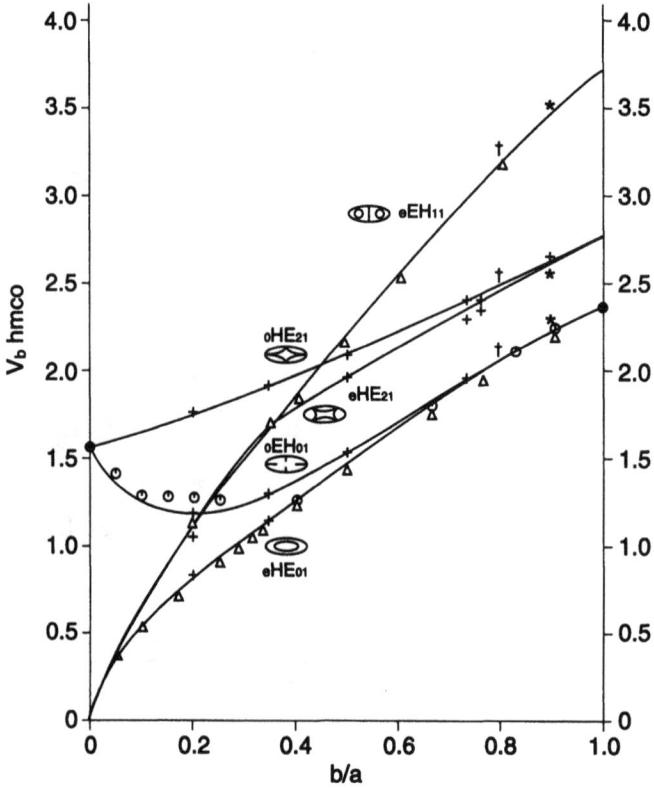

Figure 5.6 Modal cutoff for elliptical waveguide with large dielectric difference. Δ, Rengarajan [9]; ⊙, Saad [11]; ⊕, Lyubimov et al. [6]; *, Lewis and Deshpande [7]; †, Yamashita et al. [8]; +, Microwave resonator.

be seen experimentally at about $b/a = 0.75$. The now hybrid $_oEH_{01}$ cutoff splits further from that of the $_eHE_{01}$ and eventually lands at $V = \pi/2$ for $b/a = 0$ at the slab.

The HE_{21} in the dielectric guide with a circular cutoff at $V = 2.81$ (compared with the H_{21} in metal guide at 3.05) splits into $_oHE_{21}$ and $_eHE_{21}$, the odd mode descending to $V_c = \pi/2$ and the even mode to $V_c = 0$ (as does the $_oHE_{21}$ and $_eHE_{21}$ in metal guide).

Thus for both metal and dielectric guides the odd-mode cutoffs go to $\dfrac{n\pi}{2}$ at the slab.

Figure 5.8 shows the likely transformation of fields for the two cases.

Also plotted on Figure 5.6 are two points for the $_oHE_{21}$ and $_eHE_{21}$ at $\xi_o = 1$; $b/a = 0.7616$ from Lyubimov et al. [6], the earliest analysis of the elliptical dielectric guide.

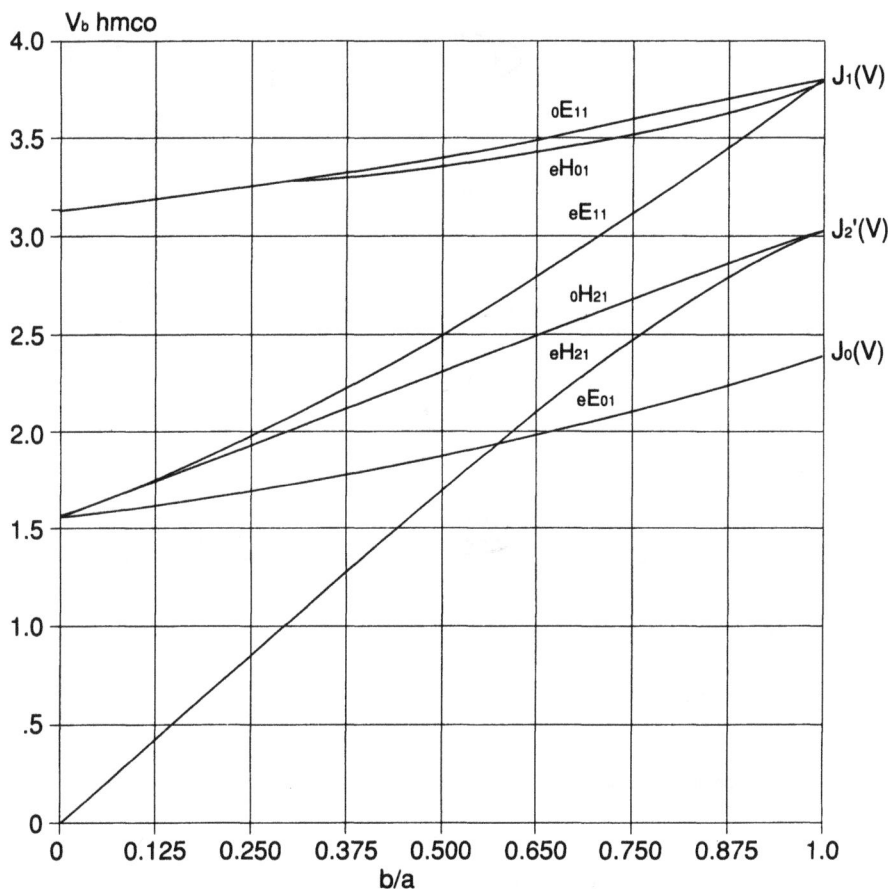

Figure 5.7 Cutoff in elliptical metal waveguide with the H_{11} lines removed.

Lyubimov's figures are for $\varepsilon_1 = 2.5$, $\varepsilon_2 = 1.0$ instead of $\varepsilon_1 = 2.3$, $\varepsilon_2 = 1.0$ for the microwave model. However, the agreement is especially good for the difference in cutoffs between $_eHE_{21}$ and $_oHE_{21}$ rather than for the actual values.

Lewis and Deshpande [7], also using the Mathieu function approach (with a 4×4 determinant, which restricts the analysis to $\xi_0 > 0.6$, $b/a > 0.54$), have calculated the cutoffs and propagation characteristics of the odd and even fundamental modes and for the next nine higher order modes for an elliptical polythene rod $\varepsilon_1 = 2.26$ in air $\varepsilon_2 = 1.0$, with $\xi_0 = 1.47$, $b/a = .90$. The cutoff values for the $_eHE_{01}$, $_oEH_{01}$, $_eHE_{21}$, $_oHE_{21}$, $_oEH_{11}$ and $_eEH_{11}$ are also plotted on Figure 5.6, showing very good agreement with the microwave experiment. As expected, no split between the $_eHE_{01}$ and $_oEH_{01}$ cutoffs is discernible at this slight degree of ellipticity, both cutoffs being at $V_b = 2.30$. However, there is a barely

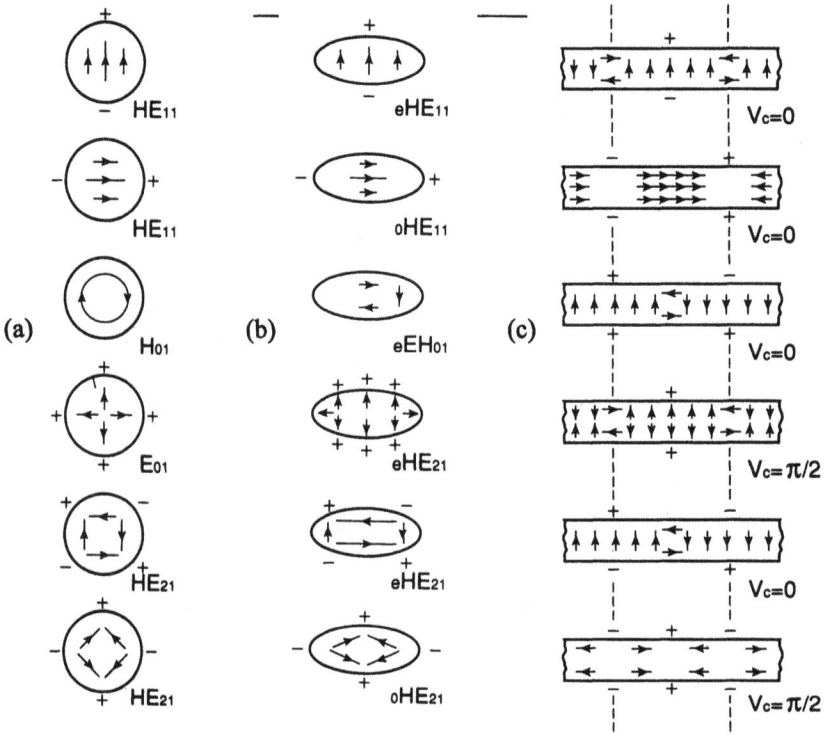

Figure 5.8 Distribution of transverse electric field for modes in (a) circular, (b) elliptical, and (c) slab dielectric waveguides.

perceptible split in the HE_{21} mode with the $_eHE_{21}$ at $V_b = 2.64$ and the $_oHE_{21}$ at $V_b = 2.69$. Yamashita et al. [8] have published cutoffs for the dielectric rod $\varepsilon_1 = 2.25$, $n_1 = 1.5$ in air for several cross-sectional shapes including the ellipse with $b/a = 0.8$ using a point-matching technique. Their values are also plotted in Figure 5.6.

Turning now to the published work on waveguides with smaller differences in index (which are good approximations for optical fibers) and considering first of all solutions in terms of Mathieu functions, Rengarajan [9] has computed the cutoffs for the first six even and six odd modes using determinants up to order 9, with $n_1 = 1.46$, $n_2 = 1.34$. The index difference $\Delta n = 0.12$ is larger than that of most practical fibers. His cutoff values for the even and odd modes (shown in Figure 5.9 (a and b)) are practically identical, with no perceptible split in the HE_{21} or $HE_{01} + EH_{01}$ modes such as those that appear in the microwave model. This implies that any such splitting starts at higher index differences. However, at $\Delta n = 0.12$ the HE_{21} modal cutoff in circular guide at $b/a = 1$ should have moved up to $V_c = 2.49$ (marked on Figure 5.9), perceptibly away from $V_c = 2.405$ at $\Delta n \rightarrow 0$.

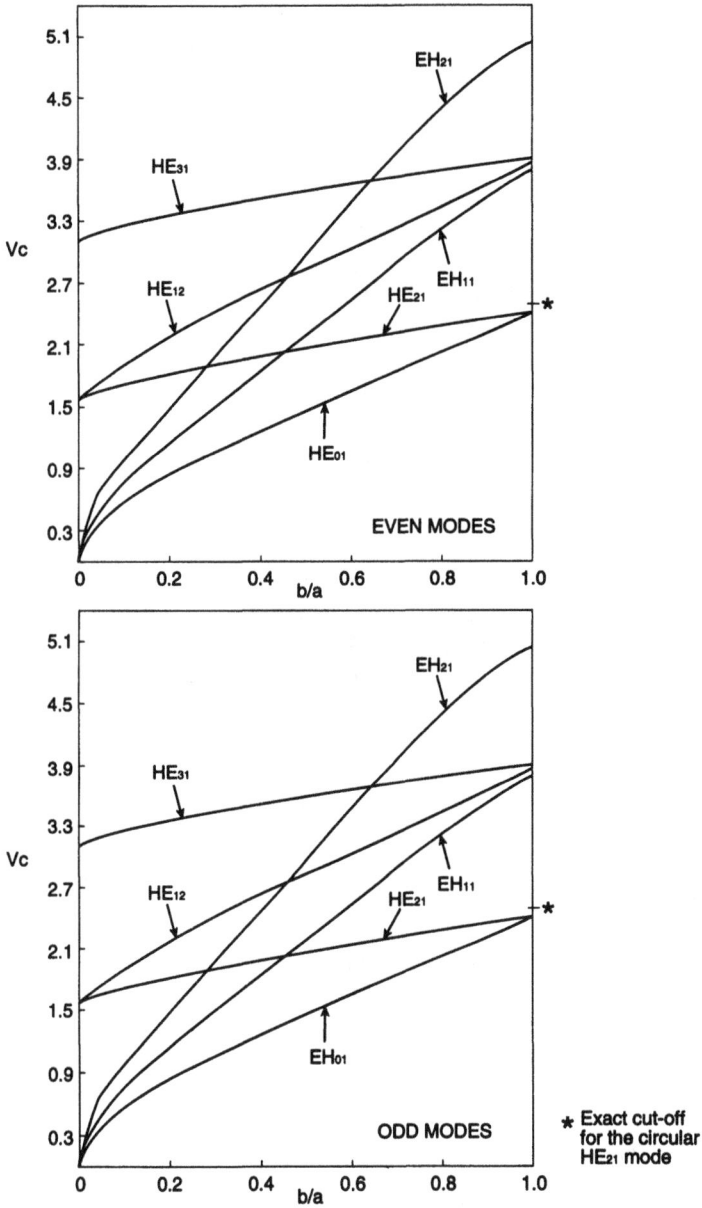

Figure 5.9 Higher mode cutoff for elliptical fibers $\Delta n = 0.12$. (From [9].)

Rengarajan's lowest cutoff line, $HE_{01} + EH_{01}$, agrees very well with that of the microwave model and with many weakly guiding analyses, suggesting that the cutoff values for the HE_{01} mode are more or less independent of index difference, as indeed are the general properties of the H_{0m} modes in circular guide [10]. These H_{0m} modes are unique in having no axial electric field, and although the elliptical version, $_eHE_{01}$, is a hybrid, the axial electric field is very weak.

This reinforces the idea that, even with small Δn, at some value of b/a the EH_{01} cutoff splits from that of the HE_{01} and ends up with $V_c = \pi/2$ at $b/a = 0$.

Evidence of such a split has been presented by Saad [11], who used a point-matching solution to the problem. A detailed description of this method applied to rect-angular guides is given by Goell [12]. Saad's analysis for identical values of $n_1 = 1.46$, $n_2 = 1.34$ shows (Figure 5.6), the EH_{01} cutoff splitting from that of the HE_{01} at about $b/a = 0.45$ and going to $V_c = \pi/2$ at $b/a = 0$. This split at $b/a = 0.45$, $\Delta n = 0.12$ compares with that at $b/a = 0.75$, $\Delta n = 0.50$ of the microwave model.

The HE_{21} splits perceptibly at $b/a = 0.9$ in the microwave model with the odd $_oHE_{21}$ mode ending up at $V = \pi/2$ and the even $_eHE_{21}$ at $V = 0$ at $b/a = 0$, but as with the $_eHE_{01}$ and $_oEH_{01}$ modes, the difference in cutoff for the two $_eHE_{21}$ and $_oHE_{21}$ modes is almost imperceptible for the weakly guiding analyses.

Eyges, Gianino, and Wintersteiner have published an extensive study of weakly guiding dielectric waveguides of various cross-sectional shapes using an integral repre-sentation technique [13]. Their values for higher mode cutoff in the elliptical guide for $b/a = 0.833$, 0.666 and 0.50 are plotted in Figure 5.10 and agree well with Rengarajan's results.

The Kumar and Varshney analysis, described in the last chapter, which uses a first-order perturbation of a rectangular section, predicts higher mode cutoffs which also agree well with the results quoted so far. They show very slight differences in the values for the even and odd modes. For instance, at $b/a = 0.5$, $n_1 = 1.485$, $n_2 = 1.470$, and $\Delta n = 0.015$, their values for modes which correspond to the $_eHE_{01}$, $_oEH_{01}$, $_eHE_{21}$, and $_oHE_{21}$ are respectively 1.5619, 1.5639, 2.2159, 2.2194.

Decotignie [14] has obtained similar results using the finite difference method with a 20×20 mesh. He points out that while the method itself cannot be used to predict higher mode cutoff directly, the values can be found to a good approximation by extending the $\bar{\beta}$ versus V_b curve to where it cuts the axis at $\bar{\beta} = n_2$. One of his points, at $b/a = 0.125$ does show some indication of the separation of the $_oEH_{01}$ mode line, moving towards $V_c = \pi/2$ at $b/a = 0$ (Fig. 5.10).

Blake et al. [15] have used a variational approach to find the core radius of an equivalent circular waveguide for each ellipticity. For the first higher order mode cutoff

$$V_b = V_{\text{cir}} \left(\frac{1 + 3\dfrac{b}{a}}{4} \right) \tag{5.6}$$

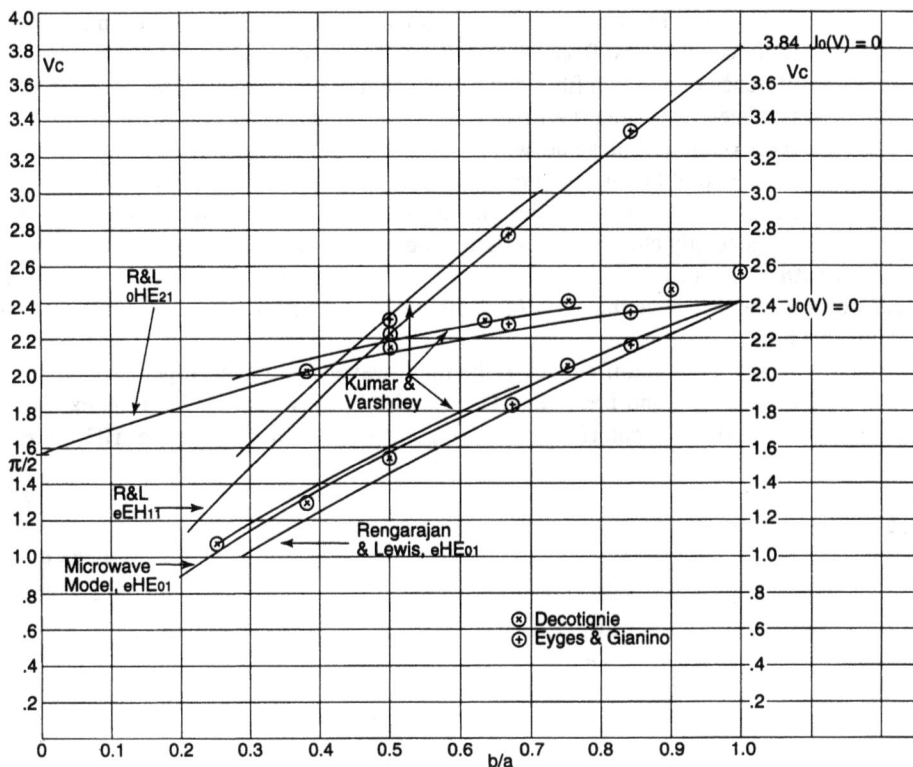

Figure 5.10 Higher mode cutoff for small index difference—comparison of values from various sources. ⊗, Points from DeCotignie [14]; ⊕, points from Eyges et al. [13].

Where V_{cir} is the first higher order mode cutoff in circular guide ($V_{cir} = 2.405$). For the second higher order mode cutoff

$$V_b = V_{cir} \left(\frac{3 + \dfrac{b}{a}}{4} \right) \tag{5.7}$$

Both these cutoff lines are shown in Figure 5.10.

An empirical expression, defined by the author, for the first higher mode cutoff is

$$V_b = 2.405 \left(\frac{b}{a} \right)^{0.6275}$$

Finally, Figure 5.11 shows some experimental points taken by the author from actual elliptically cored fibers. Because it is difficult to predict the exact core dimensions and index difference of a drawn fiber due to diffusion of the core-cladding boundary, the assumption has been made that the first higher order mode cutoff is at a value given by the microwave experiment, a value which agrees very well with the various analyses quoted above. A ratio of the higher cutoff wavelengths to the first cutoff wavelength then gives the values of V_c for those modes above the first cutoff. As production fibers, the ellipticities are generally close to the design value of $b/a = 0.5$ with a core-cladding index difference Δn of about 0.035. The core shapes are approximately elliptical, but by no means exact. There is nevertheless a fair amount of agreement with a consensus of the various theoretical analyses.

To summarize the cutoff situation, the microwave experiments together with some points by Lyubimov [6] and Lewis and Deshpande [7] show that for large index differences the $_eHE_{01}$ and $_oEH_{01}$ cutoffs split so that at the slab, $b/a = 0$, V_c for the $_eHE_{01}$ is zero

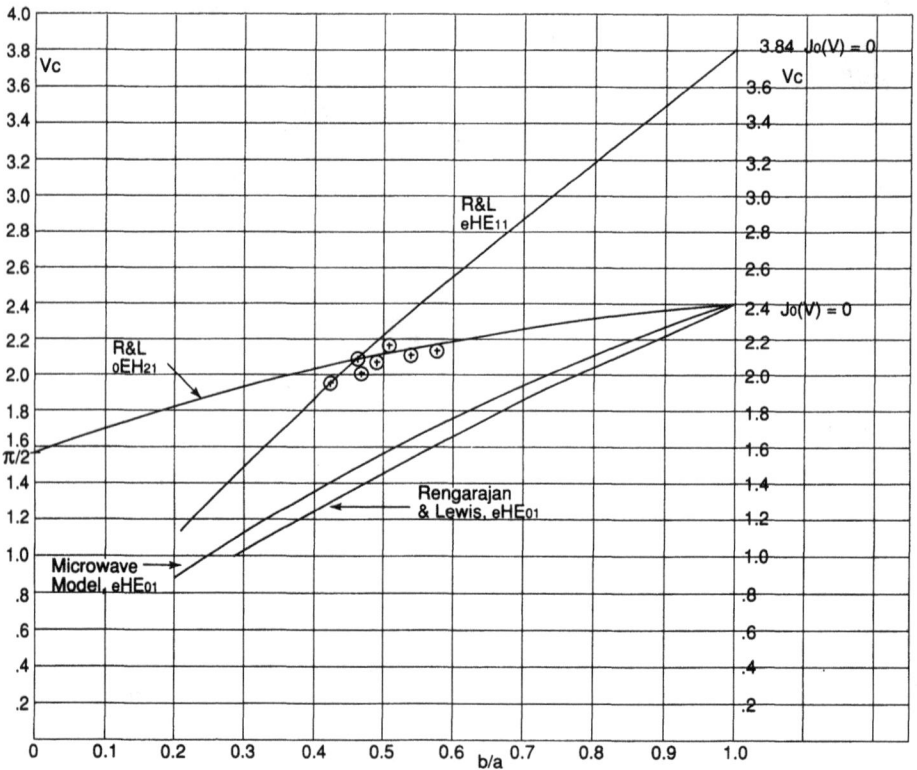

Figure 5.11 Higher mode cutoff experimental points from fiber measurements.

and for the $_o\mathrm{EH}_{01}$ is $\pi/2$. Similarly, the HE_{21} in circular guide splits into $_e\mathrm{HE}_{21}$ and $_o\mathrm{HE}_{21}$ with V_c at $b/a = 0$ being zero for the even mode and $\pi/2$ for the odd mode.

For the weakly guiding approximation and from measurements on practical fibers, there is effectively no split between either $_e\mathrm{HE}_{01}$ and $_o\mathrm{EH}_{01}$ or between $_e\mathrm{HE}_{21}$ and $_o\mathrm{HE}_{21}$ for $b/a > 0.2$. A possible explanation is that for small ellipticities and/or weak guiding, the cutoffs of each pair of modes are pulled together by the common axial fields E_z and H_z. They spring apart as the ellipticity increases; the point of separation moving to smaller ellipticities as the index difference, and therefore the guidance, of each individual mode becomes stronger. Because at the slab limit $b/a = 0$ the even modes $_e\mathrm{HE}_{01}$ and $_e\mathrm{HE}_{21}$ have cutoffs at $V = 0$ and the odd $_o\mathrm{EH}_{01}$ and $_o\mathrm{HE}_{21}$ have cutoffs at $V = \pi/2$, both sets of modes must eventually split at different values of b/a, which approach zero for zero index difference.

At approximately $b/a = 0.45$ the descending cutoff line of the EH_{11} mode cuts through the $_o\mathrm{HE}_{21}$ line, narrowing the region between first and second cutoffs at the lower values of b/a. For fibers used in such a region (e.g., for strain sensors), it makes sense to work with $b/a = 0.45$, which gives the maximum wavelength difference between cutoffs.

Between $b/a = 0.2$ and $b/a = 0$ there is a plethora of mode cutoff lines (not shown in these figures) from the EH_{1m} modes, all of which have $V_c = 0$ at $b/a = 0$ and which descend from their various cutoffs in circular guide, which are the solution to $J_{1m}(V_c) = 0$.

5.4 PROPAGATION CONSTANTS OF THE HIGHER ORDER MODES

There are few published theoretical calculations of the propagation constants for large differences in core-cladding indices.

Figure 5.12 shows the results of Lewis and Deshpande [7] for the even and odd fundamental modes and the first nine higher order modes on a polythene rod dielectric constant 2.26 (refractive index 1.50) in air with an ellipticity $b/a = .90$. Their horizontal ordinate a/λ has been converted to $V_b = \dfrac{2\pi b}{\lambda_0}[\varepsilon_1 - \varepsilon_2]^{\frac{1}{2}}$. Their vertical scale ordinate has been inverted for comparison with some results from the microwave experiment [5], shown in Figure 5.5. As b/a decreases to zero, the mode lines must eventually coalesce to form those of the slab shown in Figure 2.4.

There are more theoretical results for the weakly guiding approximation $n_1 = n_2$ and Figure 5.13(a, b) shows those of Eyges et al. [13] for two ellipticities with the horizontal ordinate converted from their $\dfrac{2b}{\lambda_0}[n_1^2 - n_2^2]^{\frac{1}{2}}$ to $V_b = \dfrac{2\pi b}{\lambda_0}[n_1^2 - n_2^2]^{\frac{1}{2}}$. The vertical ordinate remains as the square of the normalized propagation constant

$$p^2 = \frac{\bar{\beta}^2 - n_2^2}{n_1^2 - n_2^2} \tag{5.8}$$

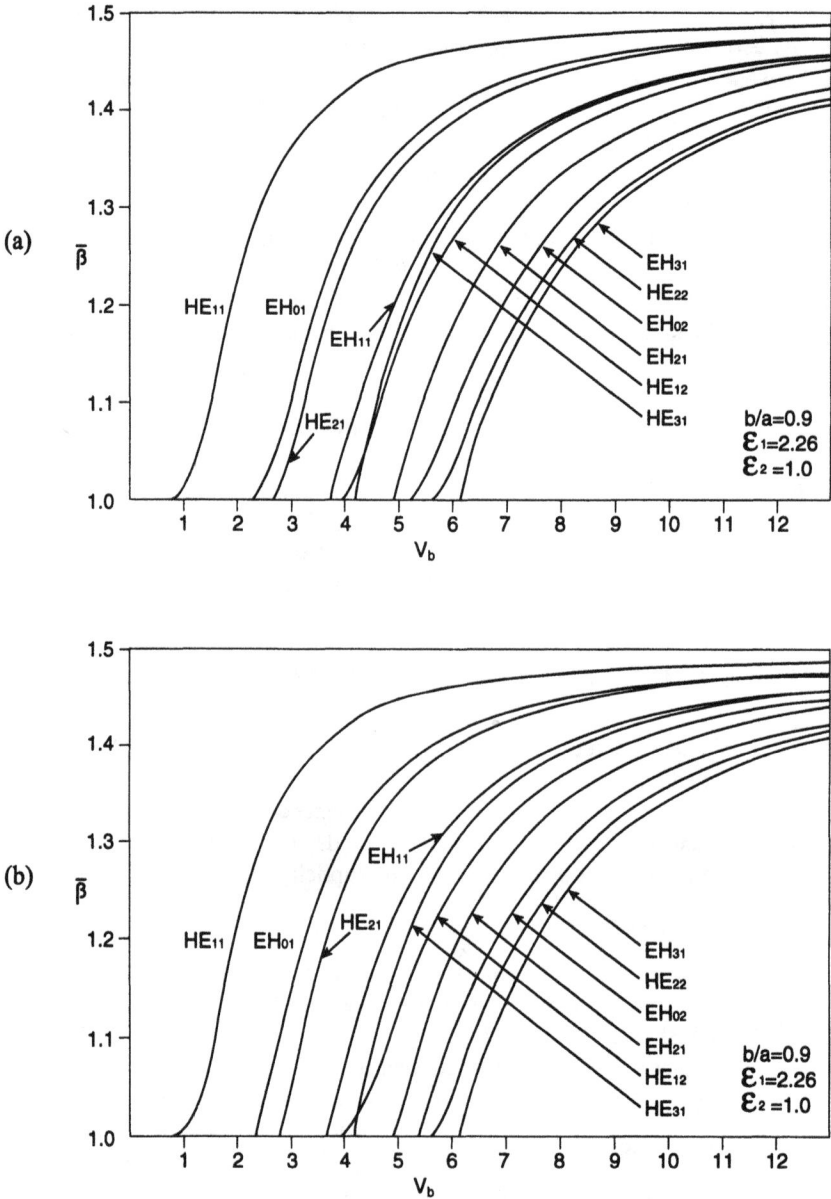

Figure 5.12 Normalized propagation constants for modes on a dielectric rod $\varepsilon_1 = 2.26$ in free space with ellipticity $b/a = 0.90$: (a) odd modes, (b) even modes. (From [7].)

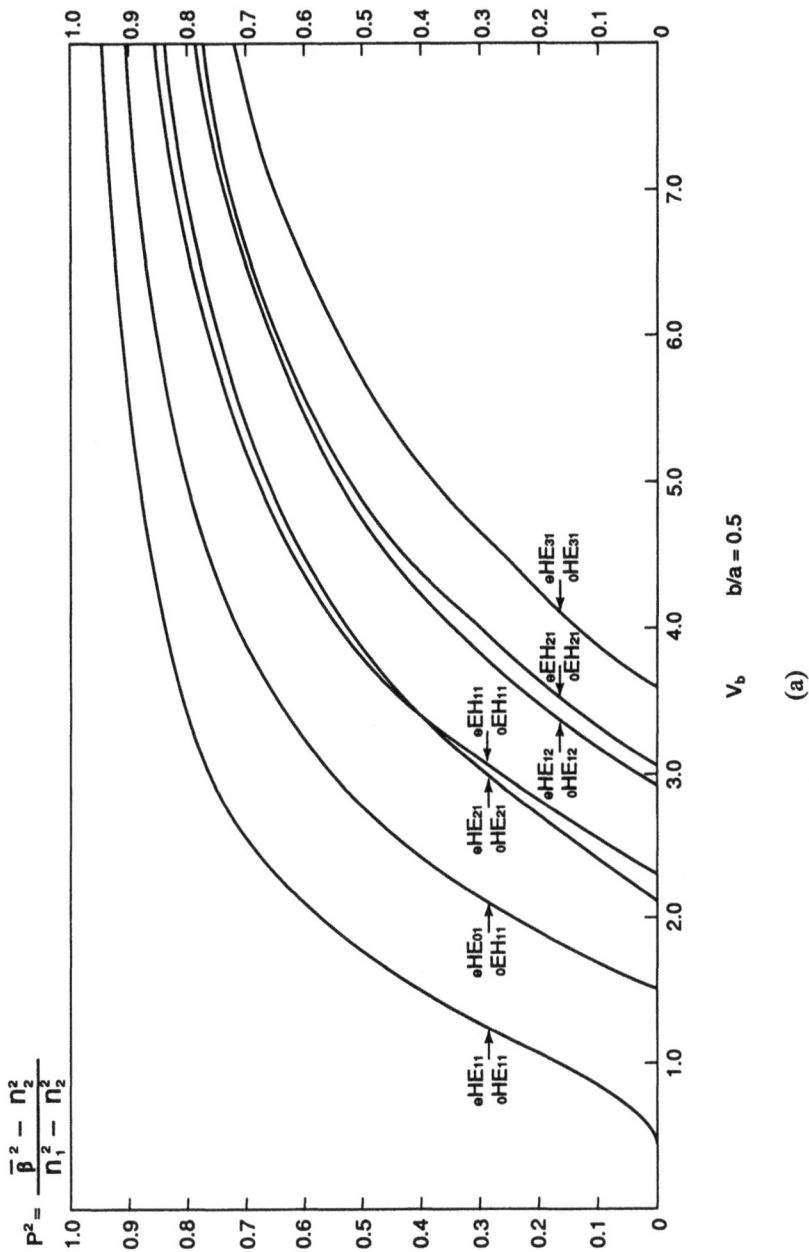

Figure 5.13 Propagation in elliptical guide with small dielectric difference: (a) $b/a = 0.5$, (b) $b/a = 0.83$. (From [13].)

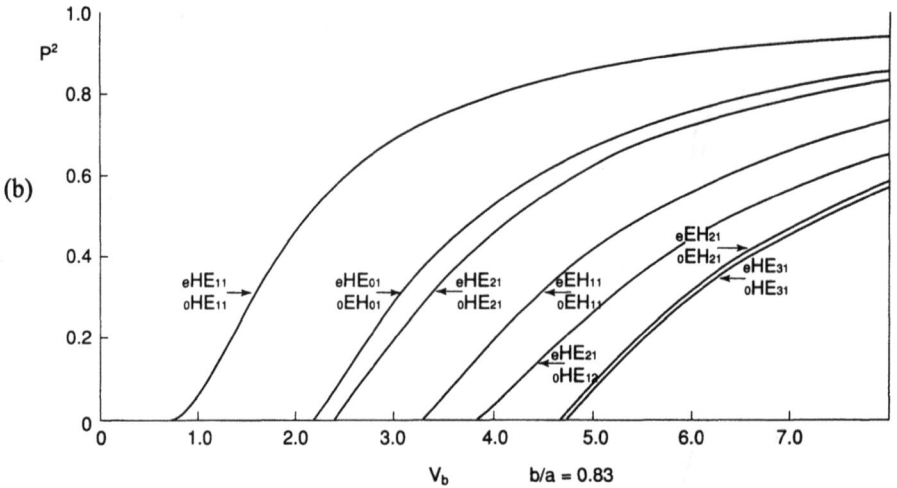

Figure 5.13 (continued).

More recently, Eguchi and Koshiba [16] have used a finite element tailored to the elliptical geometry to calculate the propagation constants and birefringence of the higher order modes for a low index difference of $\Delta n = 0.015$.

The patterns of the transverse electric fields and the power density distributions for the first few higher order modes are shown in Figure 5.14.

The circularly symmetric H_{01} and E_{01} modes in circular core fiber both have an annular pattern of power density. As the circular section becomes elliptical, the power density is concentrated in the regions of strongest fields for both $_eHE_{01}$ and $_oEH_{01}$ modes towards the tips of the ellipse. The power density splits from a single region for the fundamental $_eHE_{11}$ and $_oHE_{11}$ modes into a double region for the $_eHE_{01}$ and $_oEH_{01}$ modes along the major axis of the ellipse since this is the largest dimension of the waveguide cross section. The situation for the $_eHE_{01}$ is analogous to that of the H_{01} mode in slightly elliptical guide as treated, for instance, by Shelkunoff [17]. Any ellipticity transforms the azimuthal variation from zero to two. It is significant that an azimuthally uniform $_eHE_{01}$ mode would involve a solution in terms of a Mathieu azimuthal function s_{e0}, which does not exist. Instead, the field is described in terms of s_{e2}.

The circular HE_{21} mode has four poles and therefore must have four regions of high power density. As the section becomes elliptical, the four poles of the $_eHE_{21}$ and the four poles of the $_oHE_{21}$ superimpose to form an annular pattern of power density.

The field pattern of the $_eEH_{11}$ mode produces three regions of high power density along the major axis of the ellipse. Similarly, the $_oEH_{11}$ produces three regions along the minor axis.

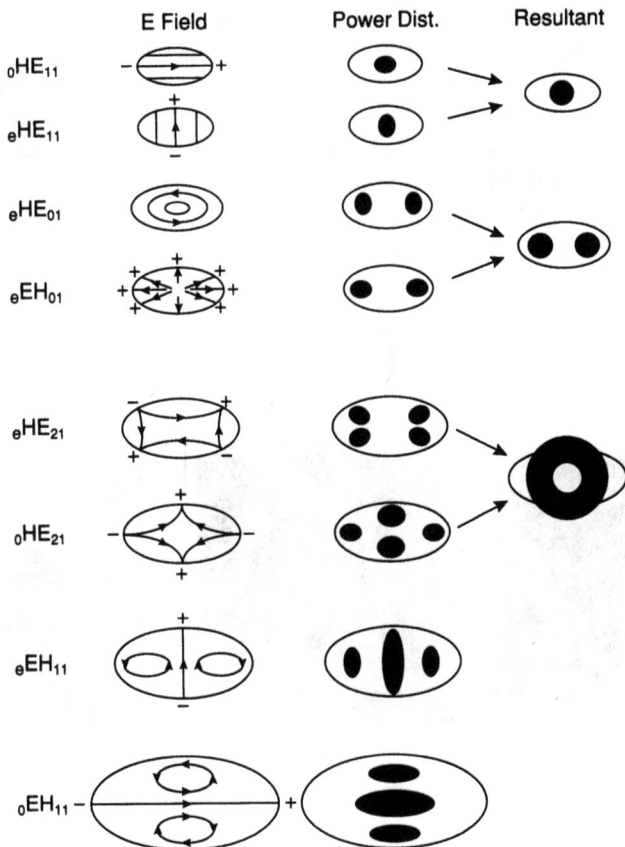

Figure 5.14 Transverse electric field patterns and power distribution.

To demonstrate these regions of high power density, light from a monochromator is launched into one end of a fiber using a high-power (×40) microscope objective. Light radiating from the other end of the fiber is projected onto the back plate of a television camera. The system is sufficiently incoherent so that there are no interference effects; that is, the fiber length is much longer than the decoherence length L_D

$$L_D = \frac{L_S}{\left[\Delta\bar{\beta} - \lambda \frac{\partial(\Delta\bar{\beta})}{\partial\lambda} \right]}$$ (5.9)

where L_S is the coherence length of the source and $\Delta\bar{\beta}$ the difference in normalized propagation constants (effective index) between the modes.

Figure 5.15 shows the patterns of light as the monochromator is tuned from long to short wavelengths. The patterns for two fibers are shown, one with small ellipticity, b/a = 0.80, and one with higher ellipticity, b/a = 0.35. As cutoff for each mode is reached, the pattern for the new mode dominates because it collects the greater amount of light from the wide-angle launching system. Comparing the patterns with the power density distributions in Figure 5.15 and referring to the cutoff diagram, Figure 5.10, the sequence of

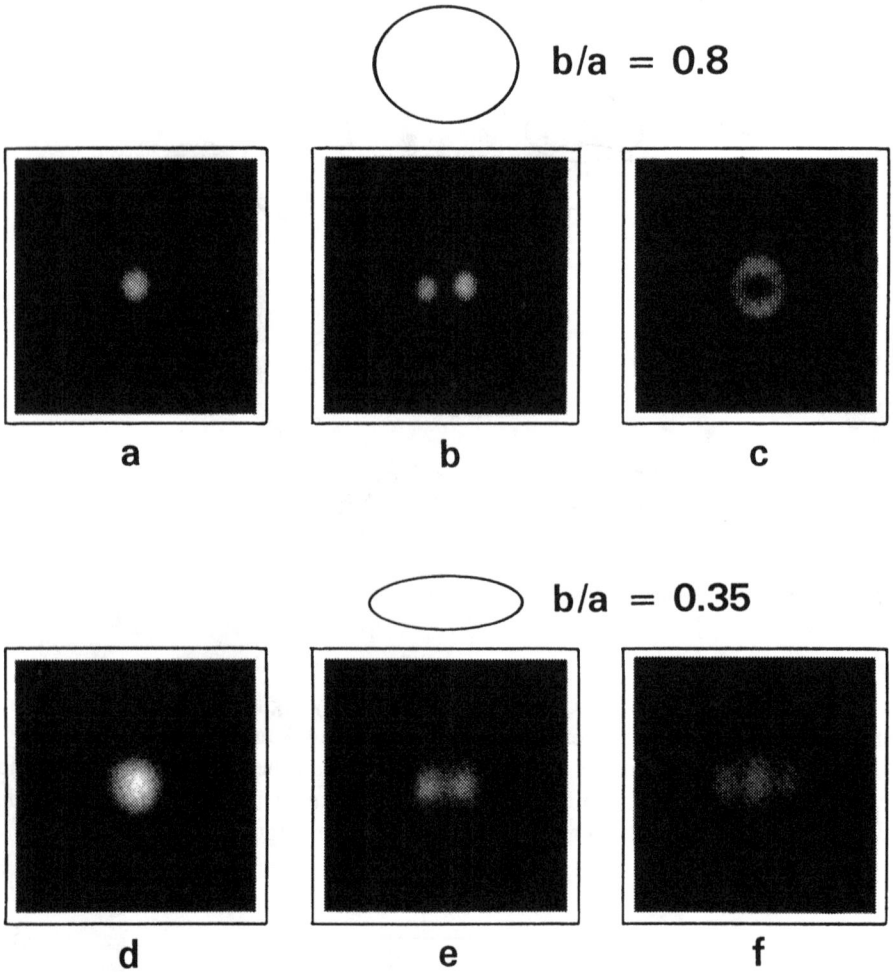

Figure 5.15 Photographs of sequences of mode patterns.

modes for $b/a = 0.80$ is (1) single region $_oHE_{11} + _eHE_{11}$, (2) two regions along the major axis of the ellipse, $_eHE_{01} + _oEH_{01}$, and (3) annular pattern formed by the four + four regions of the $_eHE_{21}$ and $_oHE_{21}$. However, for the fiber with $b/a = 0.35$, the two-region pattern $_eHE_{01} + _oEH_{01}$ is followed by the three-region pattern along the major axis characteristic of the $_eEH_{11}$ mode, confirming that (as shown in Figure 5.10) at $b/a = 0.35$ the $_eEH_{11}$ mode is the next mode to appear.

5.5 THE LP$_{11}$ MODES

For the circular core fiber, the LP$_{11}$ mode [18] is made up of the H_{01}, the E_{01} and the HE_{21}, the two former modes having the same cutoff at $V = 2.405$ with the HE_{21} cutoff slightly higher depending on the index difference Δn. However, above cutoff each mode has a propagation constant, phase velocity, and group velocity that is different from the others (see Figure 5.16(a,b)), the difference increasing with Δn. Thus, if all three modes are launched in phase into a fiber at $L = 0$ there will never be a distance L along the fiber where all three modes are once again in phase. The same applies to the elliptical fiber, but

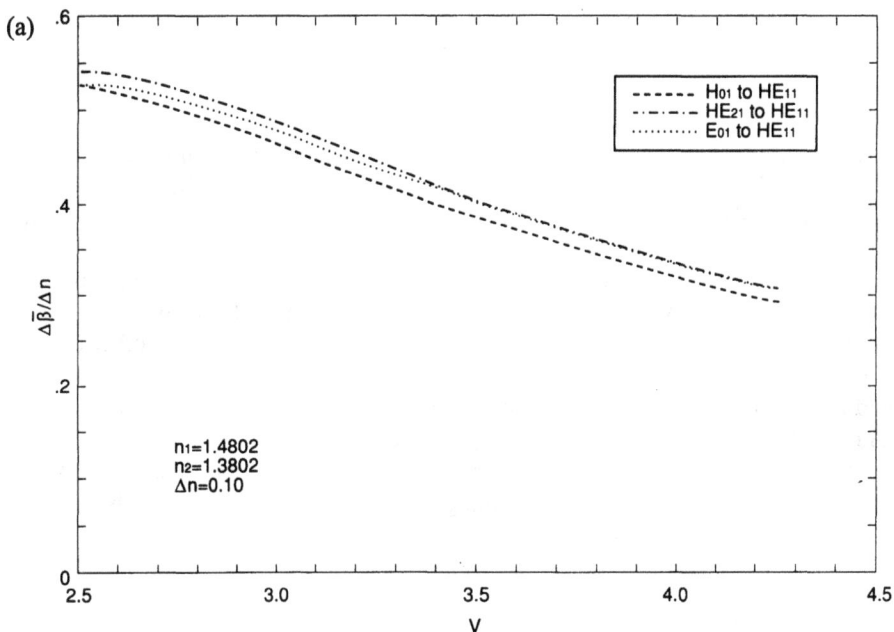

Figure 5.16 (a) Difference in normalized propagation constants in circular fiber. (b) Difference in group indices in circular fiber.

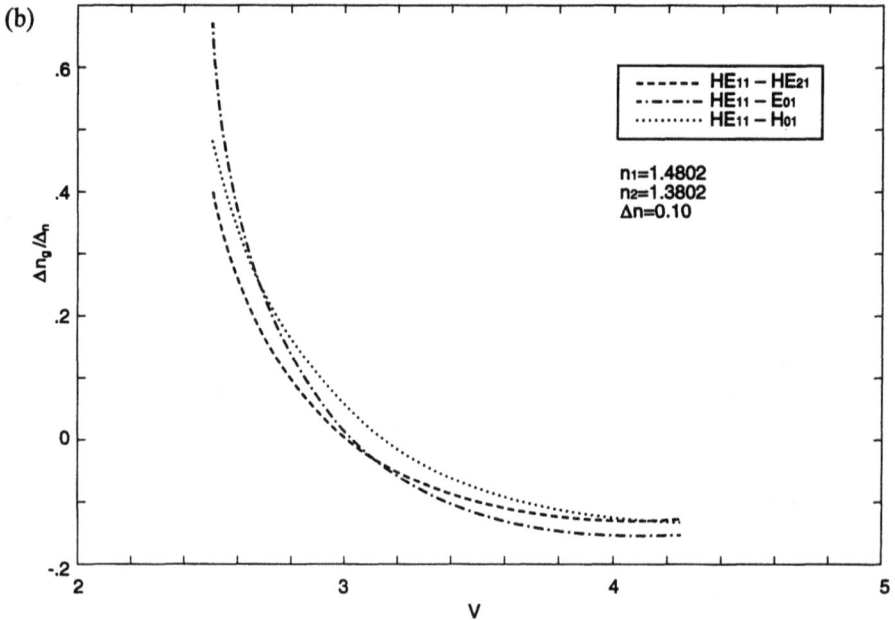

Figure 5.16 (continued)

here as the ellipticity increases there is a widening separation between the cutoffs of the $_eHE_{01}$ and $_oEH_{01}$, which are virtually identical, and those of the $_eHE_{21}$ and $_oHE_{21}$ (also virtually identical). Again, away from cutoff all four modes have different propagation characteristics. For a fiber with $0.45 < b/a < 1$, the $_eHE_{01}$ and $_oEH_{01}$ could be lumped together as the first LP_{11} mode, often called the even LP_{11} mode except that one of the constituent modes is even and the other odd. Similarly, the $_eHE_{21}$ and $_oHE_{21}$ could constitute the second or odd LP_{11} mode. Again, one mode is even and the other odd. The first higher mode cutoff would then introduce the first LP_{11} mode and the next cutoff would be that of the second LP_{11}. However, at ellipticities $b/a < 0.45$ the second cutoff heralds the EH_{11} mode in lieu of the $_eHE_{21} + _oHE_{21}$.

The LP mode situation thus becomes complicated and ambivalent and it may be simpler in the long run to abandon it altogether and to deal directly with the actual waveguide modes themselves.

5.6 BIREFRINGENCE IN THE OVERMODED REGION

One of the applications for elliptically cored fiber is as a strain gauge [19]. The fiber is operated in the higher order mode region where the beat length between the fundamental and first higher order mode propagation constants is used to measure the fiber extension.

For the circular core fiber, the higher order mode birefringence can be calculated directly from the transcendental equations. To a good approximation, the higher mode birefringence $\bar{\beta}_{HE_{11}} - \bar{\beta}_{H_{01}} = \Delta\bar{\beta}_h$ varies directly with Δn for $\Delta n < .1$ so that $\dfrac{\Delta\bar{\beta}_h}{\Delta n}$ is constant for each value of V_b.

For instance, at cutoff $V = 2.405$

with $n_1 = 1.5$

with $\Delta n = 0.1$, $\dfrac{\Delta\bar{\beta}}{\Delta n} = 0.524$

$\Delta n = 0.01$, $\dfrac{\Delta\bar{\beta}}{\Delta n} = 0.531$

For elliptical fiber, at the first cutoff

$$\bar{\beta}_{,HE_{01}} = n_2$$

It remains to find the birefringence of the elliptical fiber away from cutoff. The propagation constants of the higher order modes can be computed directly from the transcendental equation involving Mathieu functions. There are values for specific ellipticities and index differences from the approximate analyses of Kumar and Varshney [20] and of Shaw, Vengsharkar, and Claus [21]. It is also possible to obtain the equivalent circular guide by Blake's method [15] and then to solve the usual transcendental equation in Bessel functions (3.1).

Reviewing the figures for the first higher mode cutoff (the $_eHE_{01}$ alone for strong guiding and the $_eHE_{01} + _oEH_{01}$ for weak guiding), it is apparent that cutoff does not change appreciably with index difference. It is also apparent that it is the $_eHE_{01}$ mode line that is unvarying, the $_oEH_{01}$ mode line being pulled into it to an extent depending on index difference and ellipticity. The H_{01} mode in circular guide, in common with all H modes with no azimuthal variation, has

$$\frac{U}{V} = \frac{n_1^2 - \bar{\beta}^2}{n_1^2 - n_2^2} \tag{5.10}$$

independent of index (see (3.1)).

Making the assumption that, for elliptical guide also, both cutoff and $\dfrac{U}{V_b}$ for the $_eHE_{01}$ mode are virtually independent of index, it is possible to scale the propagation constant $\beta_{,HE_{01}}$ from a known value at any Δn, then to calculate $\Delta\bar{\beta}_h$ and to show that $\Delta\beta_h \propto \Delta n$ also for the weakly guiding elliptical fiber. Note that for the weakly guiding fiber $\beta_{,HE_{01}}$ becomes the propagation constant of the first $_1LP_{11}$ mode.

Figure 5.17 shows $\dfrac{\Delta\bar{\beta}_h}{\Delta n}$ against V_b for various ellipticities. On the figure are plotted

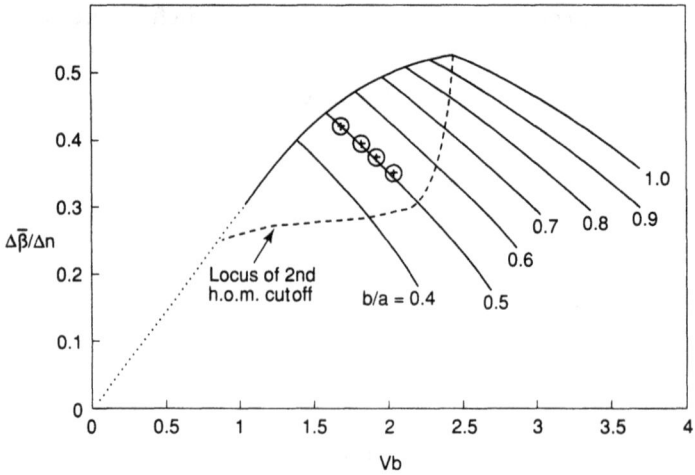

Figure 5.17 Normalized birefringence between $_eHE_{01}$ and $_oHE_{11}$ modes: $\oplus$, points from Kumar and Varshney [20].

the values at higher mode cutoff, the values at $b/a = 1.0$, and points from [20]. The solid lines are derived by scaling from microwave results.

The dashed line on the figure encloses the region between the first and second higher mode cutoffs (see Figure 5.10). The optimum operating region is at about $b/a = 0.5$, where the difference between the cutoff wavelengths is greatest.

5.7 DIFFERENCE IN GROUP VELOCITY BETWEEN THE FUNDAMENTAL AND FIRST HIGHER ORDER MODES

The group velocity v_g is given as

$$v_g = \frac{\partial \omega}{\partial \beta} \tag{5.11}$$

A more convenient term for optical fibers is the group index $n_g = \dfrac{c}{v_g}$. Again, starting with the circular guide, the LP_{11} mode is made up of the H_{01}, E_{01}, and HE_{21} modes. These weakly guiding modes have nominally the same cutoff at $V = 2.4048$, although the HE_{21} mode cutoff is slightly higher at

$$V_c \approx 2.405 + Log_e \frac{n_1}{n_2} \tag{5.12}$$

At any given value of V, the propagation constants for these three modes differ by only a few percent. For instance, with $n_1 = 1.4802$, $n_2 = 1.3802$, and $\Delta n = 0.1$, $V = 2.6$ (just above higher mode cutoff, the values of $\bar{\beta}$ for the H_{01}, E_{01}, and HE_{21} modes are 1.3823, 1.3821, and 1.3808 with the value for the fundamental HE_{11} mode at $\bar{\beta} = 1.4372$. Thus the difference in propagation constants between these three higher modes and the HE_{11} mode are 0.0549, 0.05510 and 0.05640 (practically identical). However, there is a far greater divergence in the corresponding differences in group index. Taking only the H_{01} and E_{01} modes, Figure 5.16(b) shows the differences in group index

$$\Delta n_{gH} = n_{g(HE_{11})} - n_{g(H_{01})} \tag{5.13}$$

$$\Delta n_{gE} = n_{g(HE_{11})} - n_{g(E_{01})} \tag{5.14}$$

Computations at various values of Δn have shown that $\Delta n_g/\Delta n$ is constant to within 5% for $\Delta n \leq 0.05$, sufficient for practical fibers.

For the elliptical guide, the $_eHE_{01}$ and $_oEH_{01}$ combine to give the $_1LP_{11}$ mode and $\Delta n_{g1} = n_{g(_oHE_{11})} - n_{g(_1LP_{11})}$ for the difference in group index between either the even or odd fundamental mode and that of the first $_1LP_{11}$ mode. There is also a group index difference involving the even and odd HE_{21} modes, combined as the second $_2LP_{11}$ mode $\Delta n_{g2} = n_{g(HE_{11})} - n_{g(_2LP_{11})}$.

Kumar and Varshney [20] have given results from their equivalent-rectangular analysis for $\Delta n_g = n_{g(LP_{01})odd} - n_{g(LP_{11})even}$ modes for Δn_g positive only and with $n_1 = 1.485$, $n_2 = 1.470$, $\Delta n = 0.015$ for $b/a = 0.50, 0.40, 0.333$. Their values are plotted in Figure 5.18.

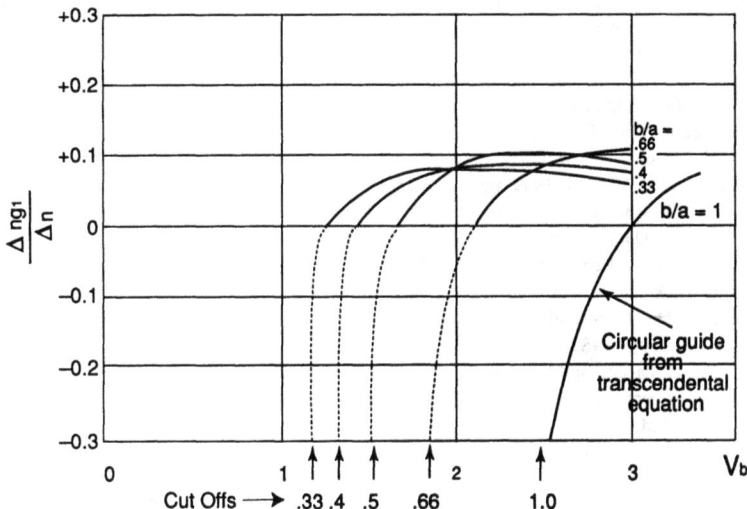

Figure 5.18 Normalized difference in group index between the $_eHE_{01}$ and $_oHE_{11}$ modes.

They point out that as the ellipticity increases, the point of zero group-index difference moves towards the higher mode cutoff. This is seen clearly in Figure 5.18.

The properties of the higher order modes in elliptical fibers (cutoff, $\bar{\beta}$, $\Delta\bar{\beta}$, n_g, and Δn_g), which in the past may have been only of academic interest, have become important as overmoded elliptical fibers have come into their own as fiber sensors.

REFERENCES

[1] Chu, L. J. "Electromagnetic waves in elliptic hollow pipes of metal," *Journal Applied Physics,* Vol. 9, 1938, pp. 583–591.

[2] Goldberg, D. H., L. J. Laslet and R. A. Rimmer. "Modes of elliptical waveguides: a correction," *IEEE Transactions on Microwave Theory and Techniques,* Vol. 38, 1990, pp. 1603–1608.

[3] Piefke, G. "Grundlagen sur Berechnung der Übertragungseigenschafter elliptische Wellenleiter," *Arch. Elekt. Ubertragung,* Vol. 18, 1964, pp. 4–8.

[4] Kretzschmar, J. G. "Wave propagation in hollow conducting elliptical waveguides," *IEEE Transactions on Microwave Theory and Techniques,* Vol. 18, 1970, pp. 547–554.

[5] Dyott, R. B. "Cut-off of the first higher order modes in elliptical dielectric waveguide: An experimental approach," *Electronics Letters,* Vol. 26, 1990, pp. 1721–1722.

[6] Lyubimov, L. A., G. I. Veselov, and N. A. Bei. "Dielectric waveguide with elliptical cross section," *Radio Engineering and Electronics (USSR),* Vol. 6, 1961, pp. 1668–1677.

[7] Lewis, J. E., and G. Deshpande. "Modes on elliptical cross-section dielectric tube waveguides," *Microwaves, Optics and Acoustics,* Vol. 3, 1979, pp. 147–155.

[8] Yamashita, E., K. Atsuki, O. Hashimoto, and K. Kamijo. "Modal analysis of homogeneous optical fibers with deformed boundaries," *IEEE Transactions on Microwave Theory and Techniques,* Vol. MTT-27, 1979, pp. 352–356.

[9] Rengarajan, S. R. "On higher order mode cut-off frequencies in elliptical step index fibers," *IEEE Transactions on Microwave Theory and Techniques,* Vol. 37, 1989, pp. 1244–1248.

[10] Snyder, A. W., and J. D. Love. *Optical Waveguide Theory,* Chapman and Hall: London & New York, 1983, p. 245.

[11] Saad, S. M. "On the higher order modes of elliptical optical fibers," *IEEE Transactions Microwave Theory and Techniques,* Vol. MTT-33, 1985, pp. 1110–1112.

[12] Goell, J. E. "A circular-harmonic computer analysis of rectangular dielectric waveguides," *Bell System Technical Journal,* Vol. 48, 1969, pp. 2133–2160.

[13] Eyges, L., P. Gianino, and P. Wintersteiner. "Modes of dielectric waveguides of arbitrary cross sectional shape," *Journal Optical Society of America,* Vol. 69, 1979, pp. 1226–1235. For a more comprehensive study, see:- In-House Report RADC-TR-79-197. 1979, Rome Air Development Center, Griffiths Air Force Base, New York.

[14] Decotignie, J. D. "Contribution a l'étude theoretique dans les guides optiques," *Thesis 463, École Polytechnique,* Lausanne, 1982.

[15] Blake, J., M. C. Pacciti, and S. L. A. Carrara. "Splitting of the second order mode cut-off wavelengths in elliptical core fibers," *IEEE Conf. Proc. 8th Optical Fiber Sensors Conference,* 1992, pp. 125–128.

[16] Eguchi, M., and M. Koshiba. "Accurate finite-element analysis of dual-mode highly elliptical core fibers," *Journal of Lightwave Technology,* Vol. 12, 1994, pp. 607–613.

[17] Shelkunoff, S. A. *Electromagnetic Waves,* Van Nostrand: New York, 1943, p. 397.

[18] Gloge, D. "Weakly guiding fibers," *Applied Optics,* Vol. 10, 1971, pp. 2252–2258.

[19] Blake, J. N., S. Y. Huang, B. Y. Kim, and H. J. Shaw. "Strain effects on highly elliptical core two-mode fibers," *Optics Letters,* Vol. 12, 1987, pp. 732–734.

[20] Kumar, A., and R. K. Varshney. "Propagation characteristics of highly elliptical core optical waveguides: a perturbation approach," *Optical and Quantum Electronics,* Vol. 16, 1984, pp. 349–354.

[21] Shaw, J. K., A. M. Vengsarkar, and R. O. Claus. "Direct numerical analysis of dual-mode elliptical-core optical fibers," *Optics Letters,* Vol. 16, 1991, pp. 135–137.

Chapter 6
Elliptically Cored Fiber:
Construction and Measurements

6.1 DESIGN REQUIREMENTS

Most of the applications of elliptically cored fibers are for fiber interferometric sensors. Here, the primary requirement is to preserve polarization by separating the propagation constants of the two fundamental modes so that light launched into one mode is not coupled into the other except by perturbations of a fairly high spatial frequency; that is, short period or tight bends. Another requirement is that any light that is launched or strays into the cladding surrounding the core is rapidly attenuated, minimizing spurious interference effects. The first requirement is met by making the birefringence as large as possible balanced against other factors such as fiber attenuation. Birefringence increases with core ellipticity and with the square of the core-cladding index difference (see Figure 4.8).

The second requirement is met by a nonguiding cladding (e.g., one with an index below that of the surrounding medium, often referred to as a depressed cladding).

A third requirement is that the fiber should be capable of being bent around a small radius of curvature without radiating. A high index difference minimizes the effect.

6.2 FIBER DESIGN

There are many ways of making circular cored fiber and most can be adapted for elliptical geometry. By way of example, one particular method will be described here.

The fiber is made by the modified chemical vapor disposition (MCVD) process with a silica body, a germania-doped silica core and a fluorine-doped silica cladding. The core-cladding index difference is $\Delta n = 0.035$ and the cladding index is depressed below that of the surrounding silica jacket by $\Delta n_c = 0.003$. The core ellipse has a ratio of major to minor axes of about two—$b/a \approx 0.5$.

The starting point of manufacture of the fiber preform is a silica tube, along the outer surface of which two opposing shallow flats have been ground. The tube is first cleaned by being heated while a mixture of oxygen and halocarbon 116 gases is passing through, effectively etching off the inner surface. The cladding is next deposited by burning a mixture of silicon tetrachloride and halocarbon 116 gases in oxygen in the tube heated to just below the softening temperature of silica. There may be several depositions forming superimposed glassed layers of fluorine-doped silica on the inner surface of the tube. The depressed cladding has an index that is 0.003 less than that of the silica tube.

The core is deposited in one pass by burning a mixture of silicon tetrachloride and germanium tetrachloride in oxygen and glassing on the deposited germania-silica powder.

Finally, the tube is collapsed under a slight vacuum to form a solid preform with an elliptical core and cladding whose major axes are perpendicular to the planes of the ground flats. Figure 6.1 shows the sections of the tube and resultant preform at the various stages of the process and Figure 6.2 is a photograph of the preform cross section.

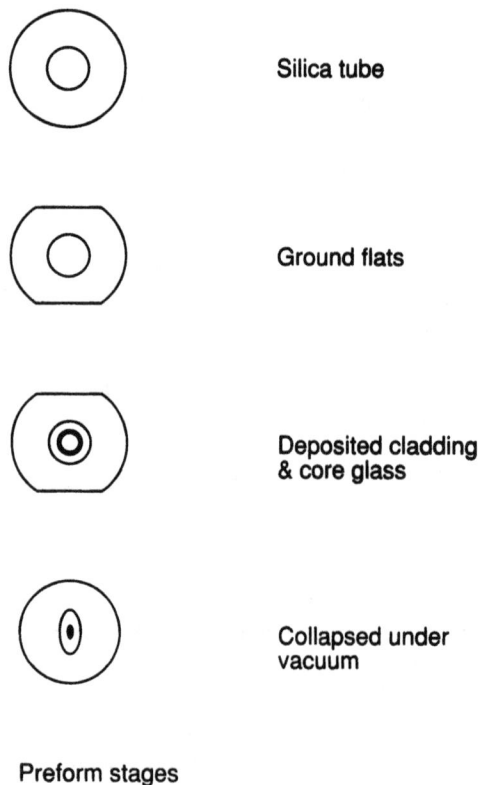

Silica tube

Ground flats

Deposited cladding & core glass

Collapsed under vacuum

Preform stages

Figure 6.1 Stages in the manufacture of the preform.

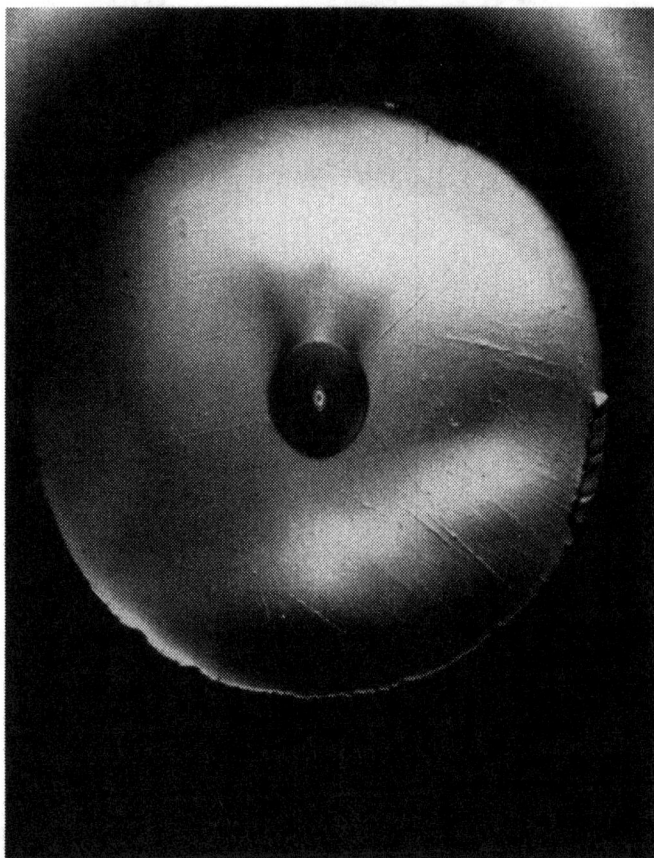

Figure 6.2 Photograph of cross section of preform.

The ellipticity of the core is controlled by the depth of the ground flats and by the degree of vacuum applied during the collapsing stage.

The preform is etched in a 10% solution of hydrofluoric acid and then flame polished to remove surface flaws. Fiber is drawn and coated at a speed of about 1 meter per second in a commercial drawing tower. During the process, the fiber is tested for tensile strength as it passes through the drawing capstan, after which it is spooled onto large diameter plastic drums. Figure 6.3 is a photograph of a fiber cross section. A range of fibers can be made to operate over wavelengths 500–1500 nm generally with outer diameters of 80 or 125m. Typically, the range or bandwidth from higher mode cutoff to the point where weak guidance begins to give increased attenuation is 50% of the operating wavelength.

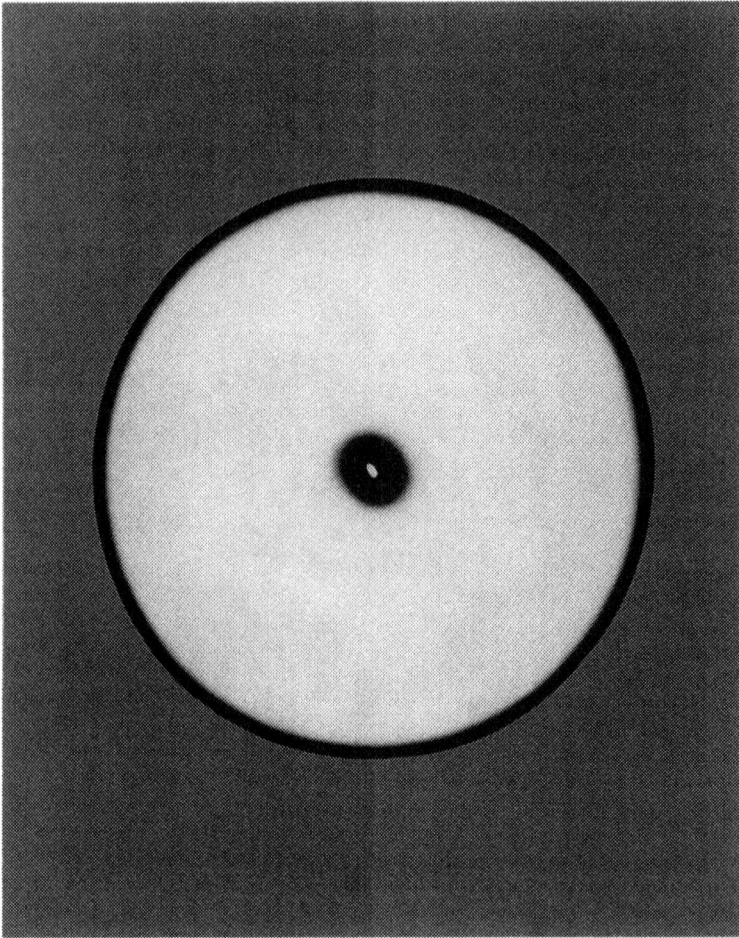

Figure 6.3 Photograph of cross section of fiber.

6.3 FIBER ATTENUATION

Fiber attenuation has two components: that due to scattering and that due to absorption. The power P reaching a distance z meters down a fiber is

$$P = P_o e^{-\alpha z} \tag{6.1}$$

where P_o is the power launched into the fiber. The attenuation coefficient α nepers/meter can be divided into two components, α_s representing loss by the scattering of light out of the fiber waveguide and α_a representing that due to absorption of the light in the waveguide material itself.

Scattering loss is caused by small inhomogeneities in the material (glass, silica), which are small compared with the wavelength. The loss therefore follows Rayleigh's Law for scattering so that α_s is proportional to λ^{-4}. Absorption loss is caused by a number of mechanisms, one of which is the presence of transition metals such as iron, chromium, and vanadium—all elements that color glass. Careful attention to the purity of the materials that go to make up the fiber can reduce this loss to negligible proportions.

Another cause of absorption loss is due to the formation of F centers when electrons in the waveguide material are displaced from their orbits by the action of light of short wavelengths in the ultraviolet (UV) region. The effect is similar to that seen in self-darkening sunglasses. It anneals itself out, the length of time to do so depending on the temperature. There is enough UV radiation from the fiber drawing furnace to produce a substantial loss in fibers that have a large concentration of germania in their composition, such as the elliptical core fiber described above. The effect can be minimized by drawing at the lowest possible temperature, thereby greatly reducing the amount of UV radiation from the furnace element, or by using a "double-draw" technique whereby fiber is given a primary draw down to an intermediate diameter and a secondary draw at lower temperature to reach its final diameter. The secondary furnace anneals the color centers induced by the UV from the primary furnace.

A third cause of absorptive attenuation is due to the OH radical, which has absorption bands at wavelengths of 0.95 µm and 1.4 µm. Meticulous attention to the dryness of the materials and of the process is necessary to keep this OH loss to a minimum. Fortunately, the chlorine released in the MCVD process is a strong drying agent in itself.

6.4 MEASUREMENT

Fiber attenuation can be measured in a number of ways, but the simplest and most generally used is the "cutback" method. Here, light from a source of variable wavelength, generally a tungsten lamp followed by a grating monochromator, is launched into a long length of the fiber. The output power over a range of wavelengths is measured using a detector (or series of detectors to cover the range). Without disturbing the launching, the fiber is cut back to about one meter and the process repeated. Figure 6.4 shows the

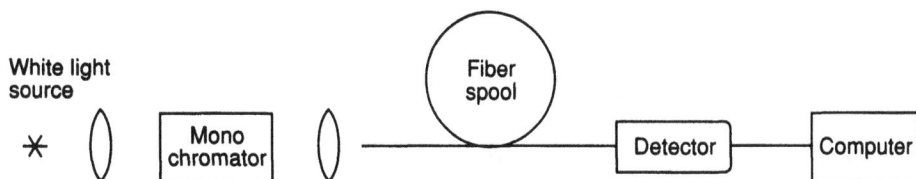

Figure 6.4 Measurement of fiber attenuation.

arrangement. The attenuation of each of the fundamental ($_e$HE$_{11}$ and $_o$HE$_{11}$) modes may be measured separately by launching light polarized along either minor or major axis with a polarizer at the output to filter out the unwanted mode.

All these measurements are best done with a chopped light source combined with a phase-sensitive detector. This gives increased sensitivity together with insensitivity to stray light that might otherwise produce spurious effects.

Figure 6.5 shows a typical attenuation plot for an elliptically cored fiber, $b/a = 0.5$ index difference $\Delta n = 0.035$ with a fluorine-doped cladding.

6.5 THE REFRACTIVE INDICES OF GERMANIA AND FLUORINE-DOPED SILICA

The effect of doping silica with germania or fluorine is to increase or decrease the index, respectively.

Sunak and Bastien [1] have used the Clausius-Mossoti interpolation scheme proposed by Melman and Davies [2] together with Malitson's data for pure silica [3] and Fleming's [4] for pure germania to obtain

$$\frac{n^2 - 1}{n^2 + 2} = (1 - f) \sum_{m=1}^{3} \frac{A_{sm}\lambda^2}{\lambda^2 - Z_{sm}^2} + f \sum_{m=1}^{3} \frac{A_{gm}\lambda^2}{\lambda^2 - Z_{gm}^2} \tag{6.2}$$

where n is the index of the composite glass, λ is the wavelength, and f is the molar fraction of the dopant.

For relatively small concentrations of germania ($< 20\%$) the germania terms (suffix "g") can be incorporated with the silica terms (suffix "s") to give

$$\frac{n^2 - 1}{n^2 + 2} = \sum_{m=1}^{3} \frac{(A_{sm} + f B_m)\lambda^2}{\lambda^2 - Z_{sm}^2} \tag{6.3}$$

The coefficients are

	$m = 1$	$m = 2$	$m = 3$
A_{sm}	0.2045154578	0.06451676258	0.1311583151
Z_{sm}	0.06130807320	0.1108859848	8.964441861
A_{gm}	0.2271125649	0.1099158881	0.1052670953
Z_{gm}	0.06092880420	0.1419148170	10.86114943
B_m	−0.1011783769	0.1778934999	−0.1064179581

with Z_m and λ expressed in μm.

For fluorine doping, the B_m coefficients in (6.3) are

	$m = 1$	$m = 2$	$m = 3$
B_{fm}	−0.05413938039	−0.1788588824	−0.07445931332

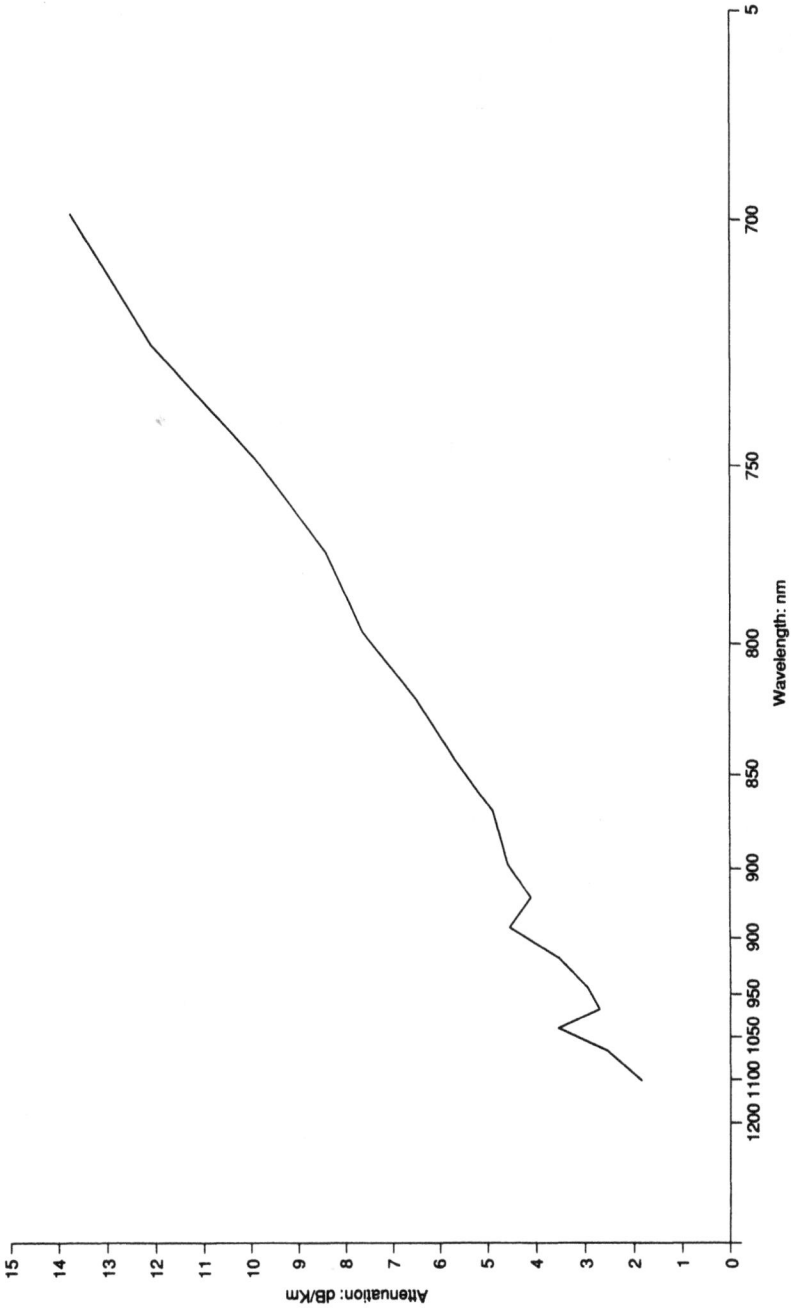

Figure 6.5 Fiber attenuation versus wavelength.

Table C.1 in Appendix C shows the index of germania-doped silica at wavelengths of 810, 1300, and 1500 nm versus molar fraction of germania.

Table C.2 shows the index and the depression of index below that of silica for fluorine doping of a few percent at the same wavelengths.

6.6 THE EFFECT OF TEMPERATURE ON THE PROPAGATION CONSTANT

An important factor affecting the stability of fiber interferometers is the variation of optical phase with temperature. The phase shift over a length L of fiber is

$$\phi = \beta L \qquad (6.4)$$

where β is the propagation constant. The normalized propagation constant (effective index of the waveguide) is

$$\bar{\beta} = \frac{\lambda_0}{2\pi} \beta \qquad (6.5)$$

so that

$$\phi = \frac{2\pi}{\lambda_0} \bar{\beta} L \qquad (6.6)$$

The change of phase with temperature T

$$\frac{\partial \phi}{\partial T} = \frac{2\pi}{\lambda_0} \left[\frac{L \partial \bar{\beta}}{\partial T} + \bar{\beta} \frac{\partial L}{\partial T} \right] = \frac{2\pi L}{\lambda_0} \left[\frac{\partial \bar{\beta}}{\partial T} + \bar{\beta}\alpha \right] \qquad (6.7)$$

The value α is the coefficient of linear expansion of the waveguide material. For silica $\alpha = 5.1 \times 10^{-7}/°C$, and $\dfrac{\partial \bar{\beta}}{\partial T}$ is approximately equal to the change in index with temperature $\dfrac{\partial n}{\partial T}$. For silica $\dfrac{\partial n}{\partial T} = 9.7 \times 10^{-6}/°C$ [5].

Thus, the change in phase with temperature is dominated by change in index rather than by change in length; for example, at a wavelength $\lambda_0 = 810$ nm with an index for silica of 1.45, and $L = 1$m

$$\frac{\partial \phi}{\partial T} = \frac{2\pi}{\lambda_0} [9.7 \times 10^{-6} + 5.1 \times 10^{-7} \times 1.45] = 81 \text{ rad/m/°C}$$

This value has been confirmed experimentally using the method illustrated in Figure 6.6. Two equal lengths of elliptically cored fiber make up the arms of a Michelson interferometer. One arm, the laser diode source, the detector, and the directional coupler are kept at room temperature in an aluminum box fixed to a metal table. The other arm is in an identical box that is heated to a stable temperature and then allowed to cool. The number of cycles of variation in detector output is noted for a given drop in temperature. The mean taken over six temperature ranges was found to be

$$\frac{\partial \phi}{\partial T} = = 82.5 \text{ rad/m/}°\text{C}$$

6.7 MEASUREMENT OF FIBER BIREFRINGENCE

6.7.1 Measurement Using Visual Observation of the Beat Pattern in the Fiber

When the light propagating in the fiber is visible and comes from a source that is sufficiently coherent, the coherence length being longer than the beat length, an accurate way of measuring birefringence is to count the number of beat lengths by viewing the light emitted from the fiber caused by Rayleigh scattering. Each scattering particle forms a radiating dipole excited by the incident light. The radiation pattern is shown with reference to the direction of the electric field of the propagating wave in Figure 6.7. Radiation is maximum orthogonal to, and zero in line with, the direction of the field.

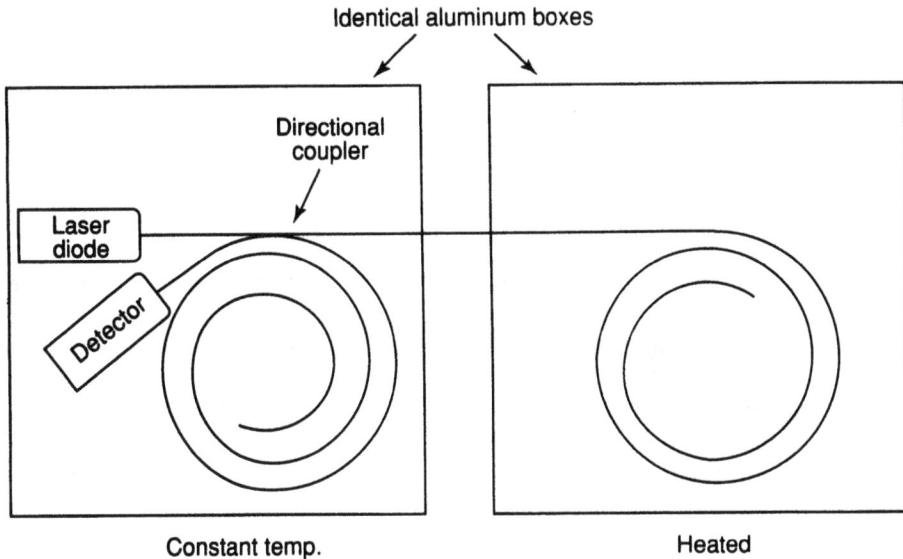

Figure 6.6 Measurement of variation of propagation constant with temperature.

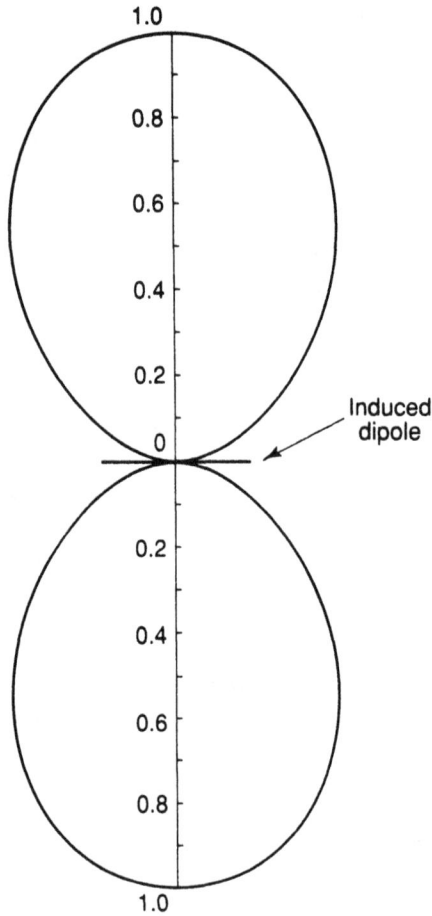

Figure 6.7 Radiation pattern from dipole.

If both the fundamental modes in the fiber are launched with equal amplitude (i.e., with the direction of polarization of the light at 45° to the birefringent axes), the polarization along the fiber progresses from linear to elliptical to circular to elliptical to linear and so on. When the fields of the waves propagating in the two modes are in phase, the radiation from the scattering dipoles is maximum, and is zero when the modes are in antiphase. The periodic light pattern can be seen by looking at the fiber at an angle of 90° to the direction of polarization of the light launched into the fiber. The arrangement is shown in Figure 6.8 where the fiber is pushed through a glass capillary tube filled with a transparent liquid such as alcohol. The capillary tube magnifies the image of the fiber making it more visible. The tube plus fiber is rotated to get the best contrast for the beat

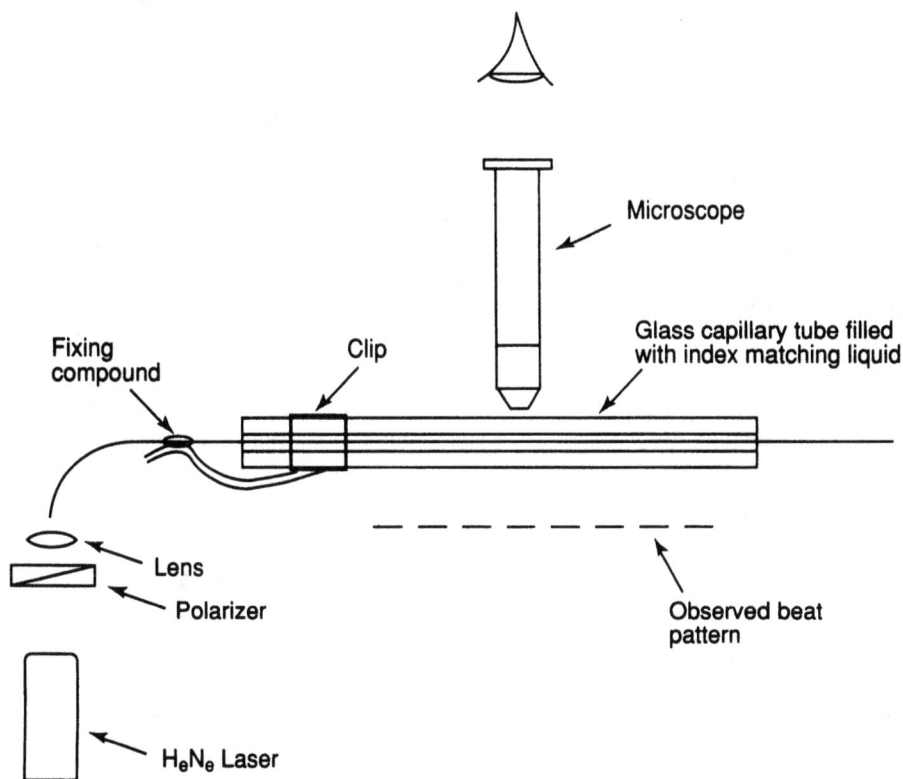

Figure 6.8 Measurement of birefringence by observing the beat pattern.

pattern. The end clips (e.g., taken from a pocket pen or pencil) ensure that the fiber rotates with the tube. A scale is fixed along the tube and the birefringence measured accurately by counting many beat lengths L_B.

Then

$$\Delta\bar{\beta} = \frac{\lambda_0}{L_B} \tag{6.8}$$

6.7.2 Measurement Using Faraday Rotation

Faraday rotation is a method originally suggested by Simon and Ulrich [6].

A magnetic field H along the axis of the fiber causes a rotation of the electric field in the propagating wave by an angle Φ

$$\Phi = VH \qquad (6.9)$$

where V is the Verdet constant, which for silica has an approximate value of 10^{-3} rad/m/gauss. Because of the difference in phase velocities of the two modes, light that is coupled from one mode to another when the fields are in phase is coupled back again when they are in antiphase. However, if the magnetic field is applied over a length that is short compared with the beat length, there will be local coupling between polarizations. By moving such a magnetic field along a fiber into which light has been launched into one of the fundamental modes and by placing a polarizer at 45° to the birefringent axes at the far end of the fiber, followed by a detector, the beat length is measured by counting the number of cycles of change in optical power over the distance the magnetic field has moved.

Figure 6.9 shows a diagram of the experimental setup. Sensitivity is greatly improved by modulating the magnetic field and using a phase sensitive detector.

6.7.3 Measurement Using Pinch Roller Method

In this method, light is again launched into one of the fundamental modes. Coupling from one mode to another is caused by local stress applied by two opposing rollers that pinch the fiber along a length that is much shorter than the beat length.

By rotating one of the rollers, fiber is pulled through in a "mangle-like" action. The beat length is measured by counting the amplitude variations over a known length of fiber passing through the rollers. Again, a chopped source and a phase-sensitive detector improves the accuracy. In a slightly different and more effective method described by Takada et al. [7], the lateral force applied at a point that moves along the fiber has a superimposed sinusoidal modulation at frequency f_m, which is used as the reference signal for the phase sensitive detector. The scheme is shown in Figure 6.10.

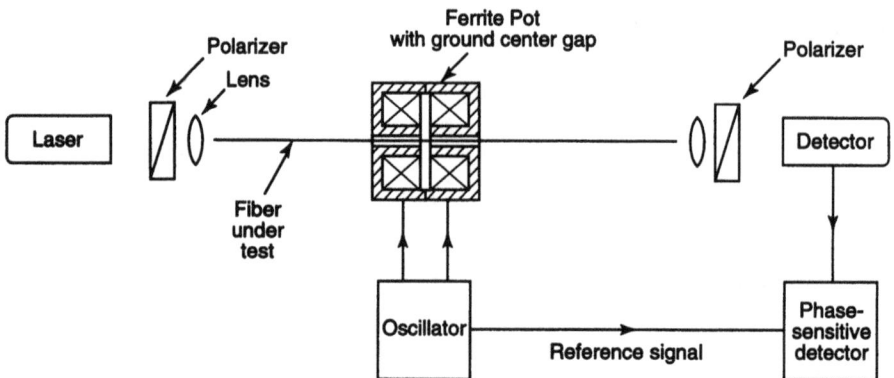

Figure 6.9 Measurement of birefringence using the Faraday effect.

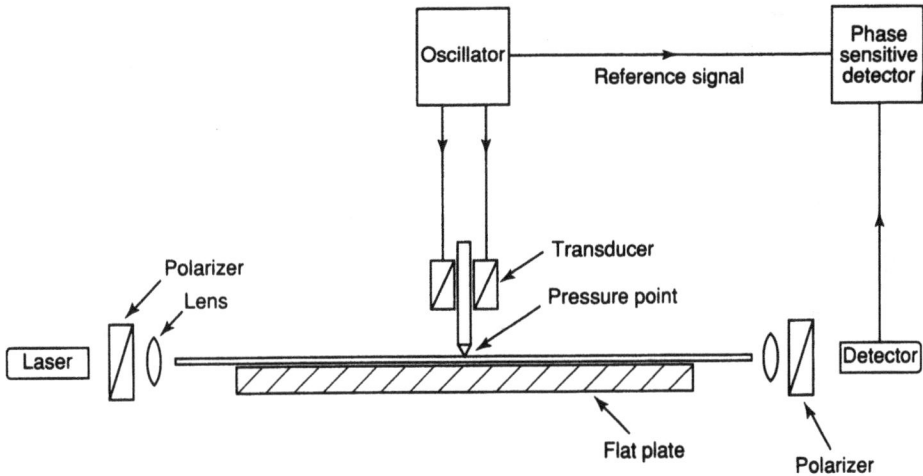

Figure 6.10 Measurement of birefringence using a pressure point.

6.8 THE EFFECT OF TEMPERATURE ON BIREFRINGENCE

The birefringence is $\Delta\bar{\beta} = \dfrac{\lambda_0}{2\pi}\Delta\beta$ where $\Delta\beta$ is the difference in propagation constants between the $_o HE_{11}$ and $_e HE_{11}$ modes. The difference in phase $\Delta\phi$ between the two modes over a distance L is then

$$\Delta\phi = \Delta\beta L = \frac{2\pi}{\lambda_0}\Delta\bar{\beta}L \tag{6.10}$$

Then the change in phase difference $\Delta\phi$ with temperature is

$$\frac{\partial(\Delta\phi)}{\partial T} = \frac{2\pi}{\lambda_0}\left[L\frac{\partial(\Delta\bar{\beta})}{\partial T} + \Delta\bar{\beta}\frac{\partial L}{\partial T}\right] \tag{6.11}$$

$$\frac{\partial L}{\partial T} = \alpha L$$

As in the previous section, the change in phase difference with temperature has two components: one due to the change in birefringence and the other to change in length, α being the thermal coefficient of linear expansion.

The birefringence $\Delta\bar{\beta}$ depends on the square of Δn (the core-cladding index differ-

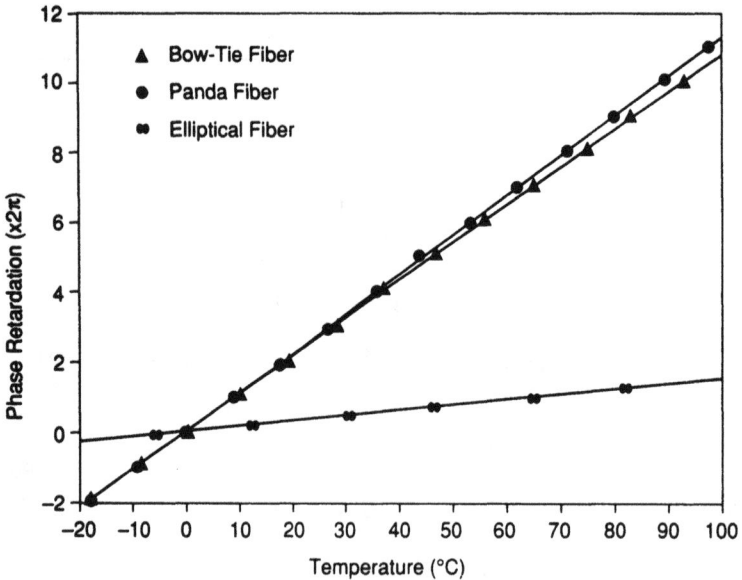

Figure 6.11 Phase shift as a function of temperature. (From [8].)

ence) and on the core ellipticity, neither of which at first sight should change with temperature. However, the difference in composition between core and cladding will cause a small variation in index difference with temperature.

A comparison has been made by Zhang and Lit [8] between the temperature sensitivity of birefringence produced by geometry and by stress, the latter effect being much greater than the former. Figure 6.11, taken from [8] shows the shift in the relative phase $\Delta\phi$ between modes at a wavelength of 633 nm in 10-cm lengths of different fibers over a temperature range of -20 to $+100°C$. The values for two fibers with stress-induced birefringence are 7.35 rad/m/°C for the bow tie fiber and 7.57 rad/m/°C for the panda fiber. For the elliptical core fiber, the effect is much smaller at 1.1 rad/m/°C. Returning to (6.11) for the elliptical fiber, the change in differential phase shift per unit length due solely to the change in length with temperature is

$$\frac{\partial(\Delta\phi)}{\partial T} = \frac{2\pi}{\lambda_0} \Delta\bar{\beta}\alpha \tag{6.12}$$

The beat length L_B for this particular fiber was measured as 1.6 mm at $\lambda_o = 633$ nm, which gives $\Delta\bar{\beta} = \dfrac{\lambda_0}{L_B} = 3.95 \times 10^{-4}$. Then, with $\alpha = 5 \times 10^{-7}/°C$ for silica, $\dfrac{\partial(\Delta\phi)}{\partial T} = 1.96 \times 10^{-3}$ rad/m/°C, which is negligible compared with the measured value of 1.1

rad/m/°C. The change in differential phase shift is therefore entirely due to the variation of birefringence with temperature

$$\frac{\partial(\Delta\phi)}{\partial T} = \frac{2\pi}{\lambda_o} \frac{\partial(\Delta\bar{\beta})}{\partial T}$$

(6.13)

Then

$$\frac{\partial(\Delta\bar{\beta})}{\partial T} = \frac{1.1 \times 633 \times 10^{-9}}{2\pi} = 1.11 \times 10^{-7}/°C$$

(6.14)

The fractional change in birefringence is

$$\frac{\partial(\Delta\bar{\beta})}{\partial T} \cdot \frac{1}{\Delta\bar{\beta}} = \frac{1.11 \times 10^{-7}}{3.95 \times 10^{-4}} = 2.8 \times 10^{-4}/°C$$

(6.15)

Since $\Delta\bar{\beta}$ is proportional to $(\Delta n)^2$ the fractional change in index difference is half the fractional change in the birefringence that it produces, so that

$$\frac{\partial(\Delta n)}{\partial T} \cdot \frac{1}{\Delta n} = 1.4 \times 10^{-4}/°C$$

(6.16)

For $\Delta n \approx 0.035$,

$$\frac{\partial(\Delta n)}{\partial T} = 1.4 \times 10^{-4}/°C \times 0.035 = 5 \times 10^{-6}$$

Thus, with $\frac{\partial n}{\partial T}$ for silica = 9.7×10^{-6}, the change in index difference necessary to produce the total change in birefringence is half the change in the index of silica alone. This implies that the change in index for the core material (approximately 0.22 mol fraction germania) differs from that of the cladding (approximately 0.015 mol fraction fluorine) by about 50%. There remains the question of sign (i.e., whether $\frac{\partial(\Delta\bar{\beta})}{\partial T}$ is positive or negative).

The value of $\frac{\partial n}{\partial T}$ (positive) for silica is greater than that of most multicomponent glasses. This is because there are two opposing effects. As the temperature increases, the index increases because of the displacement of the absorption edge in the UV towards longer wavelengths, but decreases as the glass expands, lowering the density. For silica, with its very small coefficient of expansion, the resultant decrease in index is also small so that the positive change in index dominates.

Figure 6.12, taken from the Schott catalog of optical glasses [9], shows $\frac{\partial n}{\partial T}$ as a function of wavelength for a variety of optical glasses having very different characteristics over a temperature range 20°C to 40°C. At 633 nm (the wavelength of the measurements quoted by Zhang and Lit [8]), there are only two glasses, SF58 and SF6, that have $\frac{\partial n}{\partial T}$ approaching the value for silica (10×10^{-6}), and both of these have high values of index (1.909 and 1.798, respectively), which would cause the absorption edge effect to dominate. Thus, it seems possible that a fiber core-glass with 0.22 mol fraction germania and $n = 1.488$ could have $\frac{\partial n}{\partial T}$ 50% lower than the value for silica, which would account for the observed $\frac{\partial(\Delta\bar{\beta})}{\partial T}$. However, this implies that with $\frac{\partial n}{\partial T}$ for the cladding greater than for the

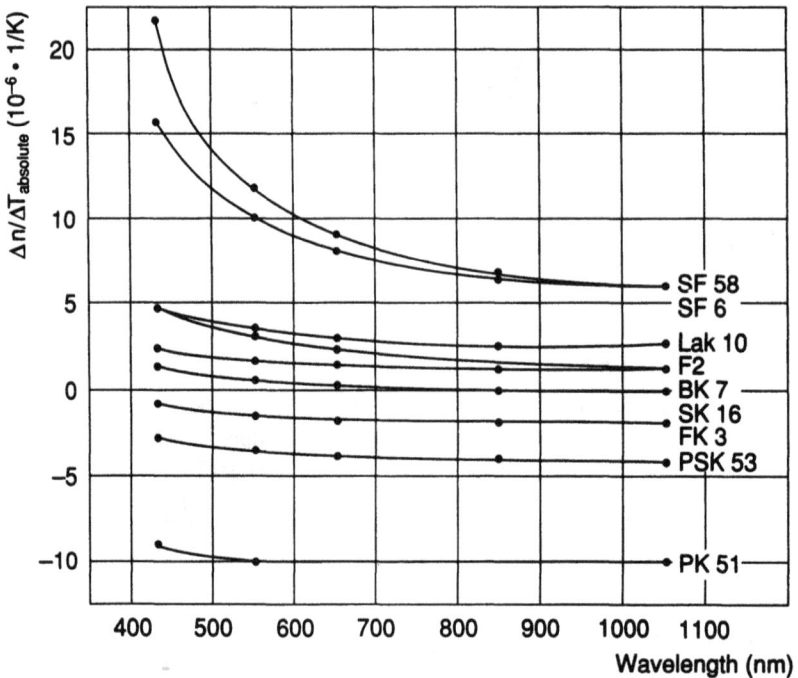

Figure 6.12 Variation of refractive index with temperature for several glasses. (From [9].)

core, Δn decreases rather than increases with temperature, with a consequent decrease in birefringence making $\dfrac{\partial(\Delta\bar{\beta})}{\partial T}$ negative.

Figure 6.13 shows an experimental arrangement designed to test this hypothesis. Light from a HeNe laser at 633 nm is launched into an elliptical core fiber that has very similar characteristics to the fiber used in [8]. The fiber is threaded through a small-diameter stainless steel tube through which a current can be passed to raise the temperature. The beat pattern is observed (as described earlier in this chapter) in the fiber emerging from the tube remote from the launched end. As the fiber is heated, the beat pattern moves away from the tube, demonstrating an increase in beat length and therefore a decrease in birefringence.

Another factor to consider is that there is a small component of stress-induced birefringence caused by the expansion mismatch between core and cladding when the fiber cools as it is being drawn and the variation in this locked-in stress with temperature changes the birefringence.

Reference [8] also describes the effect of longitudinal strain on birefringence, again comparing stress-induced and elliptical core fibers. The elongation produces changes in radial stress via Poisson's ratio.

Figure 6.14, taken from [8] shows the phase shift with elongation for 67-cm lengths of the same fibers used in the temperature tests. The quoted strain sensitivities are 107.74 rad/mm for the bow tie fiber, 102.54 rad/mm for panda fiber and 3.77 rad/mm for the elliptical core fiber, of which 1.18 rad/mm is due solely to the elongation of the fiber. This means that the birefringence due to stress in the elliptical fiber is only about 2.5% of the total.

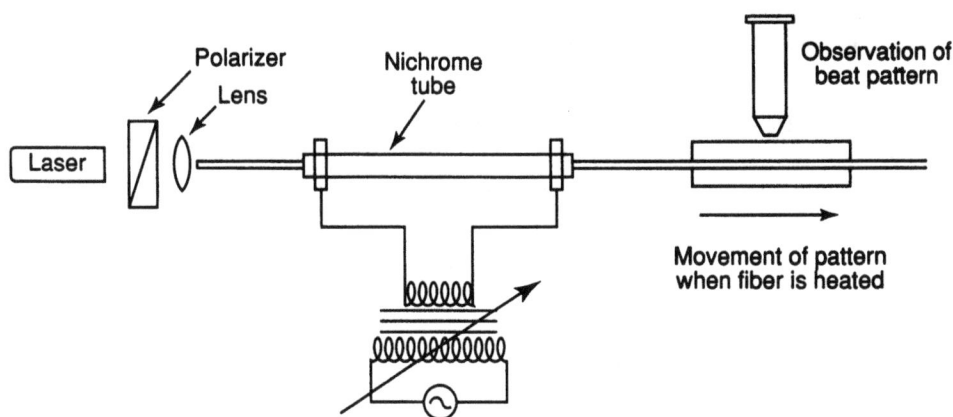

Figure 6.13 Determination of the sign of $\dfrac{\partial(\Delta\bar{\beta})}{\partial T}$.

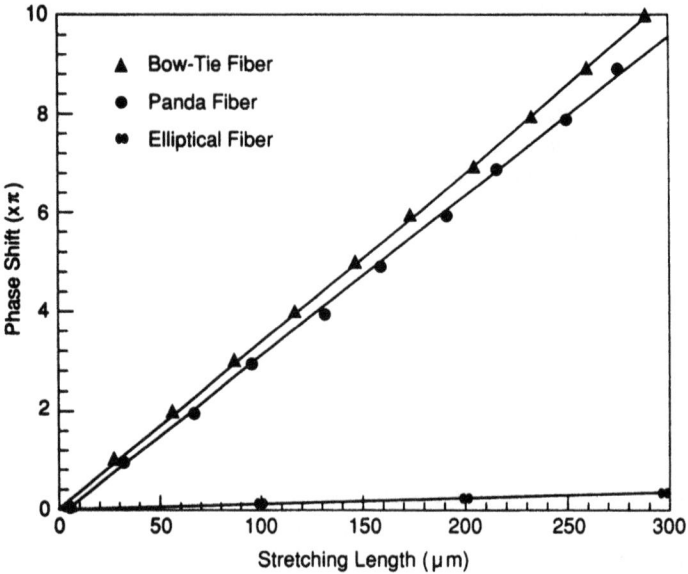

Figure 6.14 Phase shift as a function of strain. (From [8].)

Birefringence increases with strain. As the temperature rises, the strain and stress lessen, reducing birefringence so that both effects, index and stress, reinforce each other. It follows that the variation of total birefringence with temperature is mostly due to the change in index difference between core and cladding.

6.9 MEASUREMENT OF GROUP VELOCITY

The group velocity, the velocity at which energy is transmitted along a fiber, is defined as

$$v_g = \frac{\partial w}{\partial \beta} \tag{6.17}$$

with a transit time over a length L

$$t = \frac{L}{v_g} \tag{6.18}$$

Figure 6.15 shows in an exaggerated form the phase diagram of an optical fiber, including the effects of the variation of core and cladding index with frequency ω. There are several ways of measuring group velocity, each dependent on the amount and type of fiber to be measured and on the equipment available.

6.9.1 Time-Domain Pulse Method

A short pulse of light from a narrowband source at a particular wavelength is launched into the $_eHE_{11}$ or $_oHE_{11}$ mode of a fiber and the time of travel of either the pulse arriving at the far end of the fiber or of the pulse reflected from the far end back to the launching end is measured (e.g., on an oscilloscope).

This straightforward method needs a long length of fiber in order to get an accurate measurement of the transit time. However, the exact timing is difficult to measure over long lengths because the pulse spreads out or disperses as the high-frequency sidebands that go to make up the pulse travel at different velocities, an effect generally dominated by the change of index of the fiber material with wavelength (the material dispersion).

Figure 6.16 shows typical arrangements for measuring the transit time of transmitted and reflected pulses.

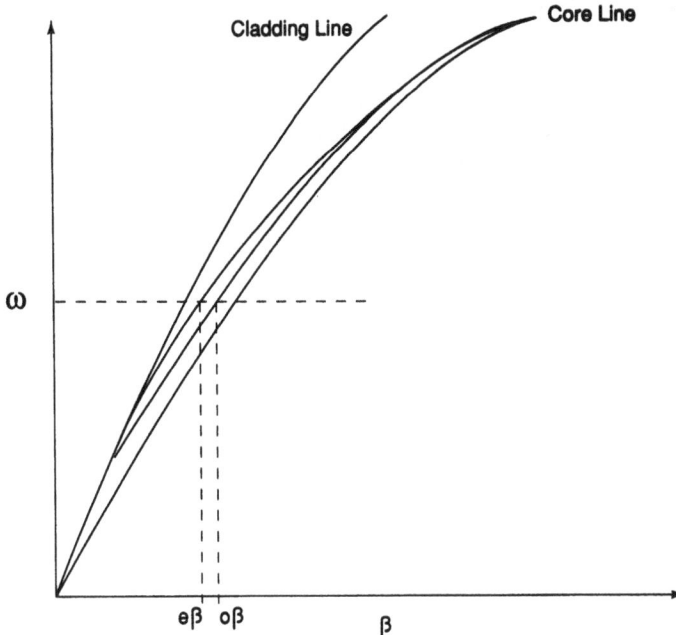

Figure 6.15 Phase diagram including variation of index with frequency.

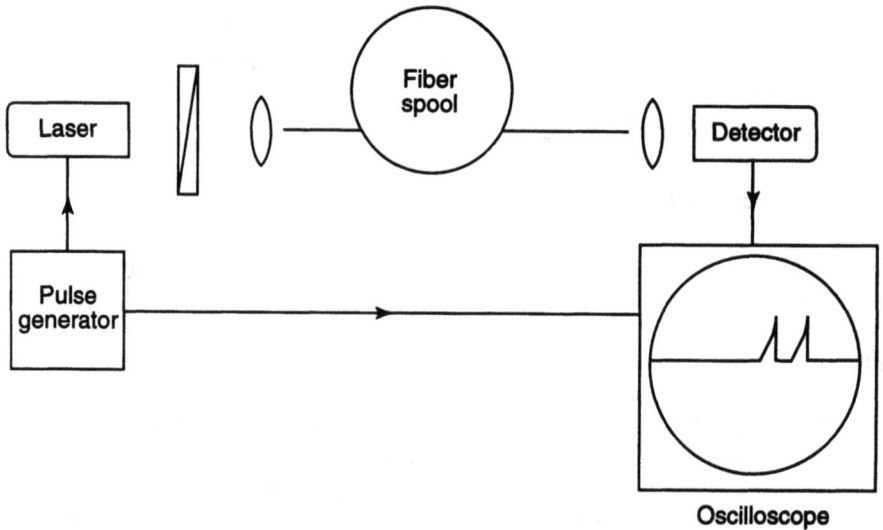

Figure 6.16 Measurement of group velocity—time-domain pulse method.

6.9.2 Frequency-Domain Modulated-Carrier Method

A coherent light source is amplitude modulated at a frequency f_m and the group velocity calculated from the phase of the modulating signal detected at the far end of the fiber. The phase shift over the length of fiber can be measured by comparing the output phase with the input phase using a vector voltmeter. Alternatively, the frequency f_m can be adjusted to give a minimum combined signal with the phase shift equal to π. The time delay t is then simply

$$t = \frac{1}{2f_m} \tag{6.19}$$

$$v_g = 2f_m L \tag{6.20}$$

This method is not affected by dispersion as only a single frequency sideband is generated, which travels at a single group velocity. Figure 6.17 shows the arrangement.

6.9.3 Resonant Cavity Method

This standard microwave technique (described in Chapter 5) for measuring the propagation constants of dielectric rods is adapted for optical fibers. The fiber itself is a resonator

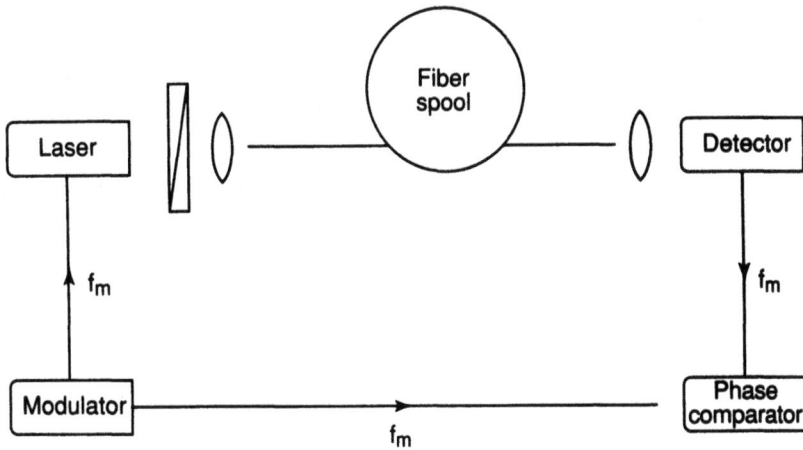

Figure 6.17 Measurement of group velocity—frequency-domain modulated carrier method.

length L with reflections at the glass-air interfaces at either end. Resonance occurs when the length is equal to an integral member N of half-guide wavelengths λ_g:

$$L = N \frac{\lambda_g}{2} \tag{6.21}$$

where

$$\lambda_g = \frac{\lambda_0}{n_e} \tag{6.22}$$

n_e = effective index of the fiber guide

In optical terminology, the system is a Fabry-Perot resonator. The transmission T as a function of free space wavelength λ_0 is given by

$$T = C \frac{e^{-\alpha L}}{1 + R^2 e^{-2\alpha L} - 2R e^{-\alpha L} \cos \dfrac{4\pi L n}{\lambda_o}} \tag{6.23}$$

where R is the reflection coefficient at the cavity ends, α is the loss in nepers/unit length, and C is a constant.

The periodicity P of the variation of transmission T with wavelength is

$$P = \frac{\lambda_0^2}{2L\left(n - \lambda\dfrac{\partial n}{\partial \lambda}\right)} \tag{6.24}$$

Pozzi et al. [10] have used this principle to find the normalized propagation constant $n_e = \overline{\beta}$ and group index

$$n_g = \left(\overline{\beta} - \lambda\frac{\partial \overline{\beta}}{\partial \lambda}\right) \tag{6.25}$$

with

$$P = \frac{\lambda_0^2}{2Ln_g} \tag{6.26}$$

In [10], a narrow linewidth laser source ($\Delta f < 200$ kHz) is tuned over a small range of wavelengths, from 1.52 to 1.58 µm, measurable to within ± 1 pm by a wavemeter of fine resolution. The detector signal gives the frequency spectrum of the transmission function as the wavelength is scanned and a Fourier analysis of the signal shows up the dominant spatial frequency corresponding to the periodicity P. Figure 6.18 shows the experimental arrangement. This method gave an accuracy (limited by the resolution of the wavemeter) of one part in 10^4 for the group index n_g.

6.10 POLARIZATION-MODE DISPERSION

The differential group delay between the $_eHE_{11}$ and $_oHE_{11}$ modes is now known conventionally as the polarization-mode dispersion (PMD), although strictly speaking the term dispersion refers to the spread in transit time caused by the variation of group velocity with frequency. It is generally treated as a nuisance in communication fibers, which have accidental birefringence caused by cores that are slightly elliptical or by bending stress.

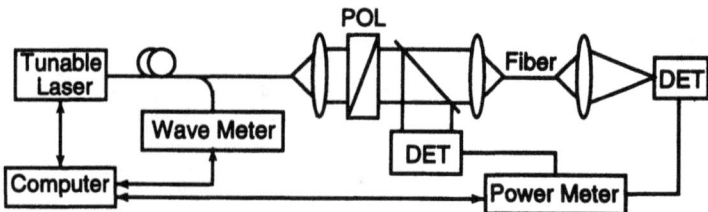

Figure 6.18 Measurement of group velocity—resonant cavity method.

For highly birefringent fibers, however, the effect is useful for various types of sensors. The differential time delay is

$$\Delta t = \frac{L}{c}\left[\Delta\bar{\beta} + V\frac{\partial(\Delta\bar{\beta})}{\partial V}\right] = \frac{L\Delta n_g}{c} \tag{6.27}$$

$$n_g = \text{group index}$$

For stress-induced birefringent (SIB) fiber, the birefringence $\Delta\bar{\beta}$ is more or less independent of wavelength (since the effective index $\bar{\beta}$ of each polarization is influenced by stress alone) and therefore the second term in the bracket can be ignored. On the other hand, elliptically cored fiber has a birefringence strongly dependent on wavelength (see Figure 4.8). At maximum birefringence, $\dfrac{\partial(\Delta\bar{\beta})}{\partial V}$ is zero, but at shorter wavelengths it goes strongly negative, decreasing the PMD to zero slightly above higher mode cutoff (see Figure 4.10). At longer wavelengths, $\dfrac{\partial(\Delta\bar{\beta})}{\partial V}$ is positive making for a PMD greater than that of an SIB fiber of the same birefringence.

PMD is measured with some of the same techniques as those used for measuring group delay.

6.10.1 Time-Domain Pulse Method

The short pulse of polarized narrowband light is launched at 45° to the elliptical axes into both, exciting both fundamental modes equally. The pulse is split into two by the PMD and the difference of arrival times measured on an oscilloscope. Again, a long length of fiber is needed to separate the pulse arrival times sufficiently for an accurate measurement, and again there is a problem with defining the pulses because of spreading due to dispersion.

6.10.2 Frequency-Domain Modulated-Carrier Method

This method is similar to that used to measure group delay with the difference that comparison of phase at the modulation frequency is now made between the signals from the two modes combined at the detector. Figure 6.19 shows the arrangement. At a modulation frequency f_{m1}, the detected signals from both modes will be in antiphase, giving a zero. The next zero will be at a frequency f_{m2} when the two signals are again in antiphase. Then,

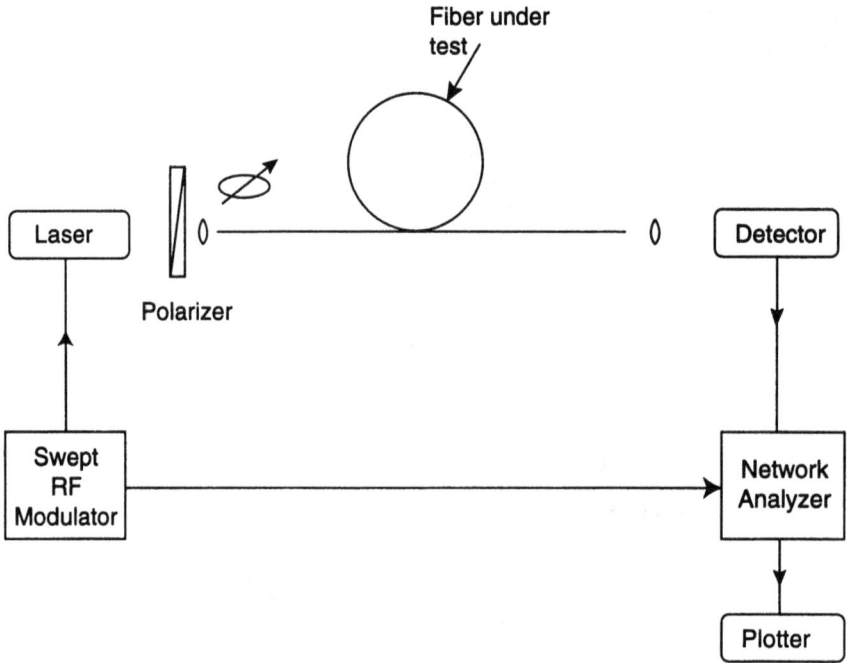

Figure 6.19 Measurement of polarization mode dispersion (PMD)—frequency-domain method.

$$f_{m1} - f_{m2} = \frac{1}{L} \frac{(v_g)^2}{\Delta v_g} \tag{6.28}$$

and the differential group delay/PMD is

$$\Delta t = L \left[\frac{1}{v_{g1}} - \frac{1}{v_{g2}} \right] \approx L \frac{\Delta v_g}{v_g^2} \tag{6.29}$$

for example, for a typical elliptically cored fiber with $b/a = 0.5$ and $\Delta n = 0.032$ operated at $V_b = 1.22$ ($V_{cutoff} = 1.56$) at a wavelength of 825 nm, the computed values of normalized group velocity $\bar{v}_g = \frac{v_g}{c}$ and normalized differential group velocity $\Delta \bar{v}_g = \frac{\Delta v_g}{c}$ are 0.676 and 1.34×10^{-4}, respectively. Then the differential time delay per unit length is

$$\Delta t = \frac{\Delta v_g}{v_g^2} = \frac{\Delta \bar{v}_g}{c \bar{v}_g^2} = 9.77 \times 10^{-13} \text{sec/m}$$

The length of fiber over which the measurement was made was 1259m, so that the calculated differential delay is

$$\Delta t = 9.77 \times 10^{-13} \times 1259 = 1.23 \text{ nsec}$$

The calculated frequency difference is then

$$\frac{1}{\Delta t} = 812.6 \text{ MHz}$$

The measured frequency difference was 852.7 MHz, corresponding to a time delay of 1.173 ns, a difference of about 5% from the calculated value.

It is possible to modulate lasers at frequencies at least up to 20 GHz, which enables much shorter fibers to be measured by this method.

6.10.3 The Modal Interferometer Method

Polarized light from a broadband source is launched into the fiber at 45° to the axes of the core ellipse, exciting both fundamental modes at more or less equal amplitudes. Another polarizer is placed at the far end of the fiber, followed by an optical spectrum analyzer (see Figure 6.20). Both polarizers are adjusted to give maximum dips in the OSA trace. The

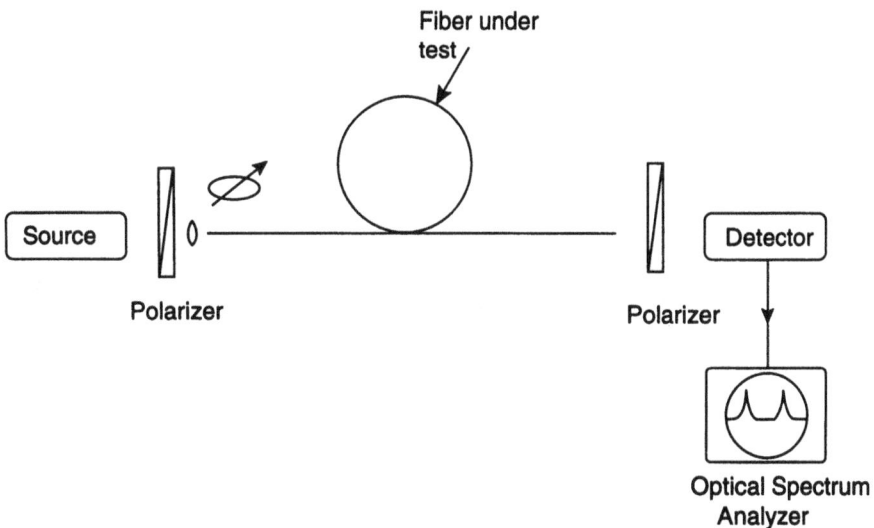

Figure 6.20 Measurement of PMD—modal interferometer method.

difference in wavelengths between adjacent minima on the trace is determined. At each minimum the light emerging from the fiber is linearly polarized in the opposite direction to the light launched into the fiber, so that between adjacent minima there has been an increase or decrease of one beat length L_B along the whole length L of the fiber. The change in birefringence $\Delta\,(\Delta\bar{\beta})$ is given by

$$\frac{\Delta(\Delta\bar{\beta})}{\Delta\bar{\beta}} = \frac{L_B}{L}$$

The differential group index is

$$\Delta n_g = \left[\Delta\bar{\beta} - \lambda\frac{\partial(\Delta\bar{\beta})}{\partial\lambda}\right] \tag{6.30}$$

so that

$$\Delta n_g = \frac{\lambda}{L_B} - \frac{\lambda^2}{L\Delta\lambda} = \lambda\left[\frac{1}{L_B} - \frac{\lambda}{L\Delta\lambda}\right] \tag{6.31}$$

The beat length L_B can be found either from a separate experiment or by cutting the fiber back by known amounts, without disturbing its azimuthal positioning, and then finding the new states of polarization emerging from the fiber (at the same wavelength).

A variation of this theme that involves the launching of three known linear states of polarization into the fiber and measuring the output polarizations at a series of wavelengths (at least two) has been devised by Heffner [11].

6.10.4 White-Light Interferometric Method

In this method Shibata et al. [12] match two paths of a Michelson interferometer to equalize the group delay/transit time so that there is interference between the waves following the two paths. Figure 6.21 shows the principle. Light from an incoherent source, such as a superluminescent diode, is launched into the fiber polarized at 45° to the birefringent axes, thus exciting the $_eHE_{11}$ and $_oHE_{11}$ modes equally. Light from the output end of the fiber is collimated by an objective lens and passes through a beam splitter to create two paths, one reflecting from a fixed mirror and the other from a mirror whose distance from the splitter is varied, generally by a motor-driven carriage.

Interference between the two paths occurs when the distances between the fixed mirror and beam splitter and the variable mirror and beam splitter are equal, and the fringes appear at the detector. Counting the number of fringes passing the detector as the variable mirror is moved calibrates the movement. Interference also occurs when the

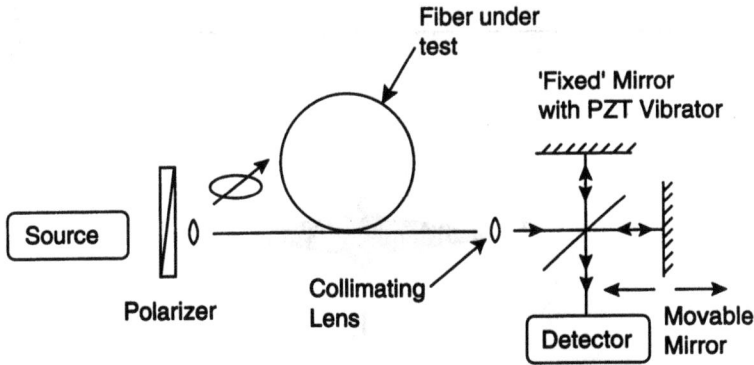

Figure 6.21 Measurement of PMD—white-light interferometric method.

difference between the fixed and variable paths creates a time delay, which compensates that between the transit time of the two modes in the fiber. If Δl is the difference in the distances between the mirrors and the beam splitter, then because of reflection the difference in the light paths is $2\Delta l$. This produces a time delay

$$\Delta t = \frac{2\Delta l}{c} \tag{6.32}$$

For a length of fiber L,

$$\Delta t = \frac{2\Delta l}{c} = L \left[\frac{1}{_o v_g} - \frac{1}{_e v_g} \right] = \frac{L}{c} [_o n_g - {_e n_g}] \tag{6.33}$$

$$\Delta n_g = {_o n_g} - {_e n_g} = \frac{2\Delta l}{L} \tag{6.34}$$

Figure 6.22 shows a plot for a typical elliptically cored fiber. This system can be made much more sensitive by vibrating the fixed mirror along the axis of the light path at a frequency f_m (for instance with a piezoelectric transducer) and then using a phase-sensitive detector with a reference signal at f_m.

This method is also very useful for finding points along a polarization-holding fiber (or along components such as a directional coupler or polarizer) where there is coupling between the modes; it is essential, for instance, in diagnosing the performance of the fiber-optic gyro. Here, light is launched into the primary mode and the coupling points show up as interference, giving the distance L_c from the coupling point

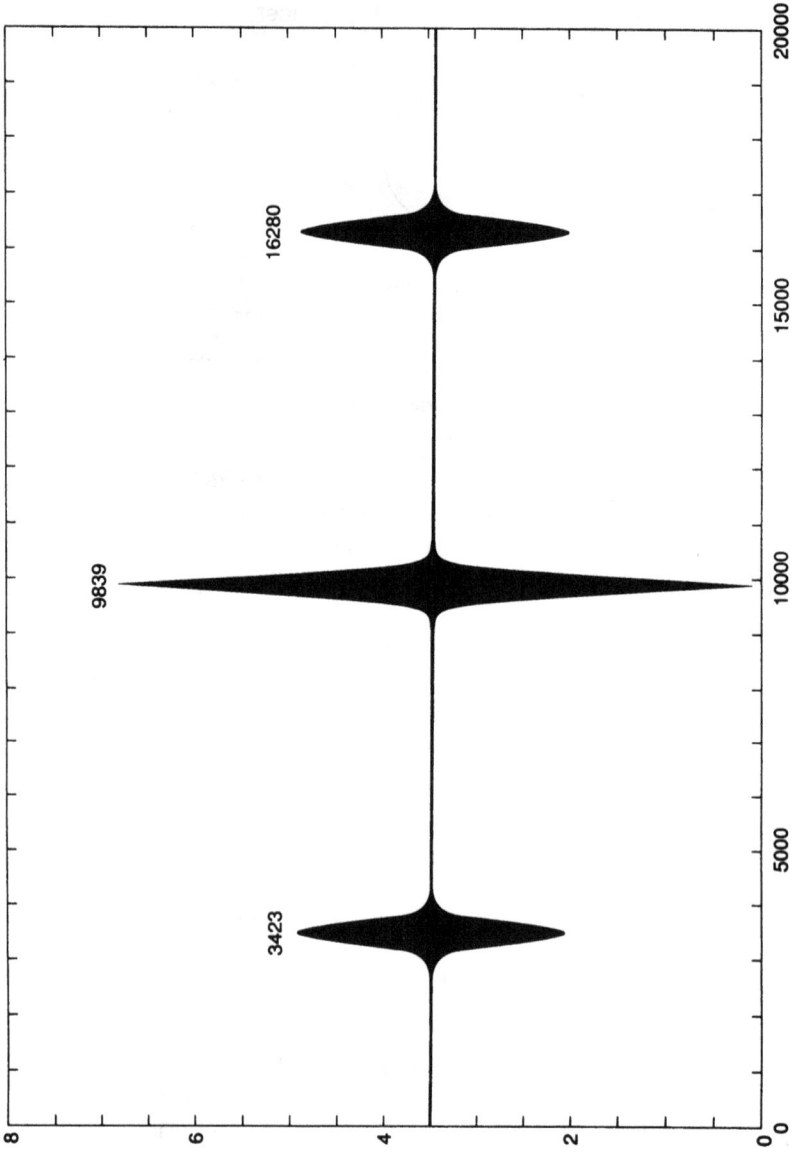

Figure 6.22 White-light interferometric plot.

$$L_c = \frac{2\Delta l}{\Delta n_g} \qquad (6.35)$$

6.11 MEASUREMENT OF THE MODE COUPLING/POLARIZATION CROSSTALK FACTOR *h*

If polarized light is used to excite only one of the principal modes in a fiber, then perturbation of the waveguide parameters will cause coupling into the orthogonal mode. The two conditions for modal coupling in any system are that fields must be generated that are common to both modes and that such fields must have a periodic component or pitch *p* so that

$$\frac{2\pi}{p} = {}_o\beta - {}_e\beta \qquad (6.36)$$

or

$$\frac{\lambda_0}{p} = \Delta\bar{\beta}; \quad p = \frac{\lambda_0}{(\Delta\bar{\beta})} \qquad (6.37)$$

That is, the pitch is equal to the "beat length" between the modes. Physically speaking, this means that power is transferred from one mode to another when both modes are in phase and when there is a field component common to both modes which does the coupling.

It is unlikely that there will be perturbation with a pitch exactly equal to and in phase with the beat length over any appreciable distance along the fiber. Such conditions would cause rapid transfer of energy between modes. However, any discontinuity (a single pressure point, for instance) can be expressed as a spectrum of spatial frequencies with some component at the beat pitch. Bending the fiber also causes modal coupling, as demonstrated in Figure 6.23.

The polarization-preserving or extinction coefficient η_e of a fiber is defined as the ratio of the power P_y in the unexcited mode to that in the excited mode P_x taken at the fiber output. Since the coupling is random along the fiber and may change with time, it is impossible to analyze the situation taking phase into account. Marcuse's mode-coupling analysis adapted by Kaminow [13] treats the coupling as random with

$$\langle P_x \rangle = e^{-hz} \cosh(hz) \qquad (6.38)$$

$$\langle P_y \rangle = e^{-hz} \sinh(hz) \qquad (6.39)$$

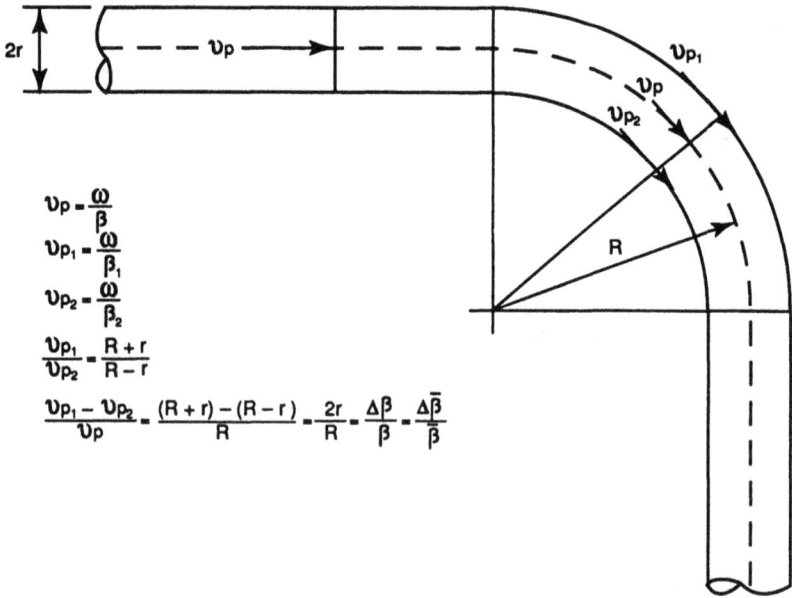

$$v_p = \frac{\omega}{\beta}$$

$$v_{p_1} = \frac{\omega}{\beta_1}$$

$$v_{p_2} = \frac{\omega}{\beta_2}$$

$$\frac{v_{p_1}}{v_{p_2}} = \frac{R+r}{R-r}$$

$$\frac{v_{p_1} - v_{p_2}}{v_p} = \frac{(R+r)-(R-r)}{R} = \frac{2r}{R} = \frac{\Delta\beta}{\beta} = \frac{\Delta\bar{\beta}}{\bar{\beta}}$$

Figure 6.23 Mode coupling due to bending.

$$\eta_e = \tanh(hz) \tag{6.40}$$

where z is the distance along the fiber and h is the coupling factor. In practice, η_e is small and $\tanh(hz) \approx hz$. Typically, in a good polarization-preserving fiber $h < 3 \times 10^{-6}$m. It is useful to express h in dB/m (as a positive value); for example, $h_{dB} = 10 \log_{10}\left(\frac{1}{3 \times 10^{-6}}\right) = 55.23$ dB. Then, $10 \log_{10}(hz) = 10 \log_{10}(h) + 10 \log_{10}(z)$ so that 55 dB/m gives a polarization extinction $\eta_e = 25$ dB at the end of 1 km of fiber.

6.11.1 Method of Measurement

The setup for measurement is shown in Figure 6.24. The light source must be sufficiently broadband to avoid coherent coupling effects in the fiber. Suitable sources are light-emitting diodes (LED), superluminescent diodes (SLD), or a white-light source with a bandpass filter. The power level of any light with a wavelength shorter than that of the higher mode cutoff must be less than -30 dB down on the total power to avoid spurious effects due to the propagation of light in the higher order modes. Light from the source is collimated and passes through a prism polarizer, generally of the Glan-Thompson type, having an extinction between two crossed polarizers of about 60 dB. The collimated polarized beam is then focused onto the well-cleaved input end of the fiber under test

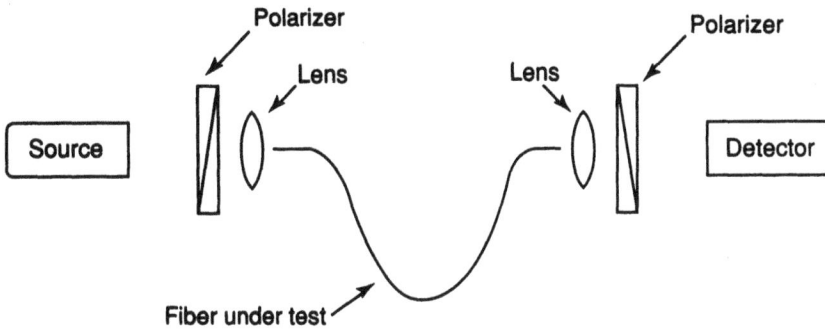

Figure 6.24 Measurement of polarization coupling.

using a strain-free lens of about ×20 magnification. The fiber should be held in position without applying so much stress that its polarization properties are affected.

Light from the well-cleaved output end of the fiber is collimated by a strain-free lens and passes through another prism polarizer onto a detector.

Care must be taken to align the output optics so that there is no residual cross-polar component of light (due to the curvature of the modal fields inside the fiber) reaching the detector. A synchronous detection system using a modulated or mechanically chopped source greatly improves the sensitivity.

The measurement is straightforward. The input and output polarizing prisms are rotated in turn until the detected signal is a minimum. The output polarizer is then rotated through 90° to give a maximum signal P_{max}. The polarization extinction ratio $\dfrac{P_{min}}{P_{max}}$ and the fiber length L_m are used to calculate h dB/m.

$$h_{dB \cdot m} = 10 \log_{10} \left(\frac{P_{min}}{P_{max}} \right) + 10 \log_{10}(L_m) \tag{6.41}$$

It is better to rotate the output polarizer. Moving the input polarizer might alter the critical alignment of the launching. A check that the light from the source is sufficiently incoherent can be made by gently moving a part of the fiber while the signal is minimum. There should be no change in signal level.

Figure 6.25 illustrates a method of finding degradation in polarization holding in a fiber due to bending. The two ends of a length of fiber are pushed through the neck of a plastic funnel to form a loop as shown. The fiber is connected to the h measuring apparatus and the polarizers crossed to obtain a minimum as described above. The fibers are then pulled down the funnel, decreasing the diameter of the loop until the signal lifts off the minimum. The bend radius is found from the position of the loop in the funnel.

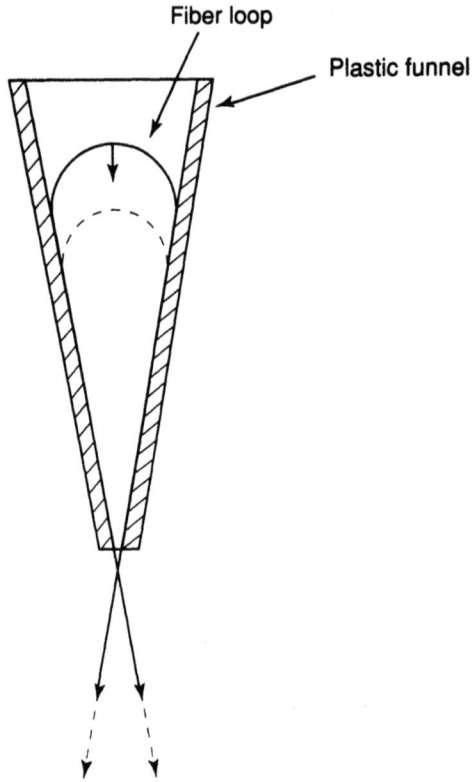

Figure 6.25 Measurement of degradation of polarization holding due to bending.

This principle can be used to measure bending loss and also fiber stress at breaking [14].

6.12 MEASUREMENT OF HIGHER MODE CUTOFF

Because of the relatively large difference in index between core and cladding in elliptical core fibers, the cutoff wavelengths for the higher order modes are well defined compared with those in weakly guiding communication fibers or fibers where the birefringence is created by stress. The point of cutoff is easily found by observing the near or far-field pattern of light emitted from the fiber; the slightly elliptical pattern divides into two along the major axis at the onset of the first higher order modes (see Chapter 5).

An arrangement for measuring cutoff is shown in Figure 6.26. A white-light source

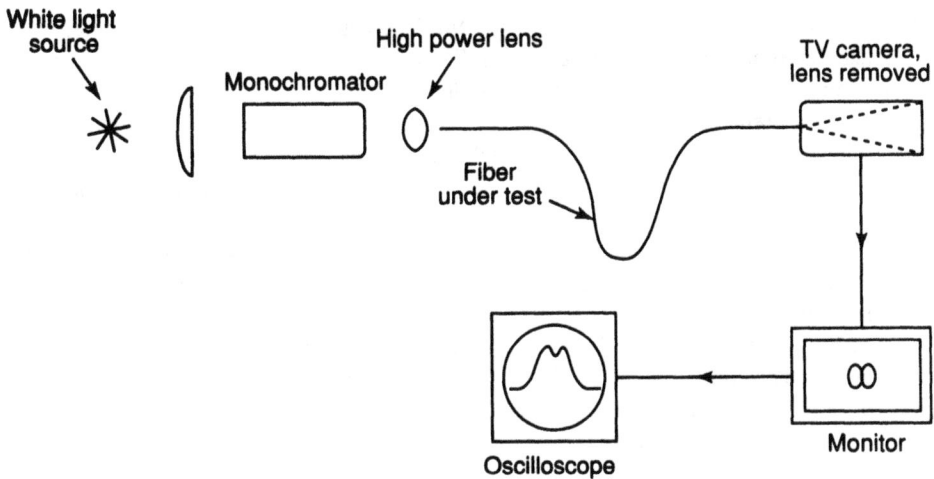

Figure 6.26 Measurement of higher mode cutoff.

is collimated to pass through a grating monochromator. The light emerging from the output slit of the monochromator is focused by a high-power (×40) lens onto the input end of the fiber to be tested.

The source is sufficiently broadband so that there is no interference between modes, that is, the fiber is longer than the decoherence length L_D

$$L_D = \frac{L_S}{\left(\Delta\bar{\beta} - \lambda\dfrac{\partial(\Delta\bar{\beta})}{\partial\lambda} \right)} \tag{6.42}$$

where L_S is the source coherence length $L_S = \dfrac{\lambda^2}{\Delta\lambda}$ and $\Delta\lambda$ is the wavelength spread. The output end of the fiber is pushed through a slightly bent metal tube (e.g., stainless hypodermic), which is fixed into a cap covering the lens opening on a TV monitoring camera, the lens itself having been removed. The far-field pattern of light radiated from the fiber thus falls directly onto the plate of the camera. The camera output is displayed on a monitor screen and is also connected to an oscilloscope with the trace synchronized to the camera line frequency. The wavelength of the light is gradually changed via the monochromator. At the first higher order mode cutoff, the pattern on the screen changes from a single patch into two patches, or vice versa. At the same time, the oscilloscope trace changes from a single peak to a double peak, or vice versa. As cutoff for each mode is reached, the pattern for the new mode dominates because it collects the greater amount of light from the wide-angle launching system. The actual transition can be determined quite accurately using both observations on TV monitor and oscilloscope. The fiber end is then removed from the camera and coupled to an optical spectrum analyzer for an accurate measurement of the wavelength. The process is repeated for the second higher order mode and so on. The cutoff wavelengths can be measured to an absolute error of less than 5 nm. This technique can also be used to measure the mode field dimensions of fibers.

A more accurate method for measuring cutoff has been described by Alavie and Grossman [15]. The arrangement is shown in Figure 6.27. The source is a tunable Ti-sapphire laser with a resolution of 0.01 nm. The fiber is bonded onto a cantilever beam and strained by loading the beam. In the multimode region, the changes in the propagation constants of the modes caused by the strain produce changes in the multilobe radiation pattern. In the single-mode region, there are no such changes. This method is reported to measure the first modal cutoff wavelength to an accuracy of 0.5 nm.

6.13 DETERMINATION OF THE AXES OF BIREFRINGENCE AND THE AXES OF THE CORE ELLIPSE

It is often necessary to know the positions of the birefringent axes of the fiber and also of the major and minor axes of the ellipse. These two sets of axes should theoretically be

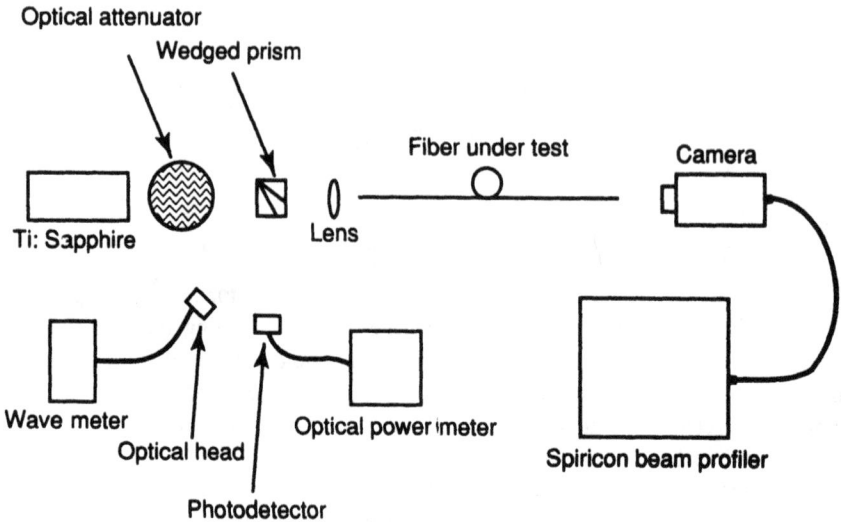

Figure 6.27 A more accurate measurement of higher mode cutoff. (From [15].)

coincident, but there may be a small misalignment either due to stress or to asymmetry of the elliptical core. If it is possible to view the fiber end-on, the birefringent axes can be found by launching light at a wavelength shorter than that at higher mode cutoff and viewing the radiation pattern from the end of the fiber. For wavelengths between the first and second LP_{11} modes, two patches of light will appear on the major birefringent axis. The light can be visible (e.g., that from a HeNe laser at 633 nm) if the wavelength falls between the two cutoffs; otherwise, an infrared viewer can be used.

If it is impossible or inconvenient to view the fiber end (for instance, when fusion-splicing), the major and minor axes of the core ellipse can be found as follows.

The method is shown diagrammatically in Figure 6.28. A beam of light from a visible coherent source (such as that from a HeNe laser at 633 nm) passes through a small hole (just large enough to allow the beam to go through) in a white screen. The beam is projected onto the fiber orthogonally to the fiber axis. An interference pattern, reflected from the elliptical core, appears on screen. When the fiber is rotated continuously about its axis, the pattern moves first in one direction, stops, and then moves back in the other direction. At the stationary point, either the major or the minor axis of the core ellipse is in line with the incident beam. The pattern moves more rapidly when the major axis is in line with the beam, making it possible to distinguish between the axes [16].

For aligning fibers during splicing, the laser beam can be split by a prism so that a separate beam is projected onto each fiber and the patterns adjusted to match exactly.

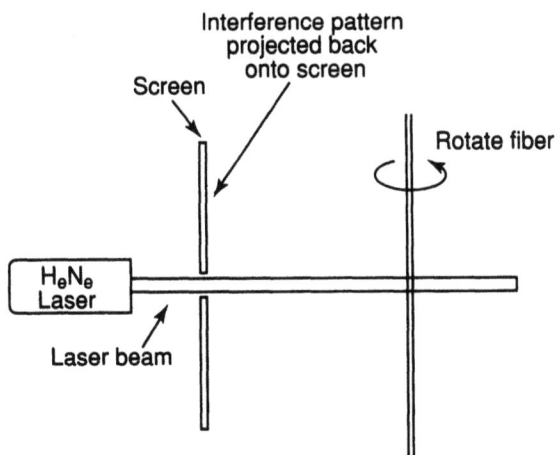

Figure 6.28 Method of finding the axes of the elliptical core in a fiber.

REFERENCES

[1] Sunak, H.R.D., and S.P. Bastien. "Refractive index and material dispersion interpolation of doped silica in the 0.6–1.8 μm wavelength region," *Proc. SPIE*, Vol. 1176, 1989, pp. 184–188.

[2] Melman, P., and R. W. Davies. "Application of the Clausius-Mossotti equation to dispersion calculations in optical fibers," *Journal Lightwave Technology*, Vol. LT-3, 1985, pp. 1123–1124.

[3] Malitson, I. H. "Interspecimen comparison of refractive index of fused silica," *Journal Optical Society of America*, Vol. 55, 1965, pp. 1205–1207.

[4] Fleming, J. W. "Dispersion in $G_eO_2 - S_iO_2$ glasses," *Applied Optics*, Vol. 24, 1985, pp. 4486–4493.

[5] Matsuoka, J. "Temperature dependence of the refractive index of silica," *Journal Non-Crystaline Solids*, Vol. 135, 1991.

[6] Simon, A. and R. Ulrich. "Evolution of polarization along a single mode fiber," *Applied Physics Letters*, Vol. 31, 1977, pp. 517–520.

[7] Takada, K., J. Hoda, and R. Ulrich. "Precision-measurement of modal bi-refringence of highly be-refringent fibers by periodic lateral force," *Applied Optics*, Vol. 24, 1985, pp. 4387–4391.

[8] Zhang, F., and J. W. Lit. "Temperature and strain sensitivity measurements of high-bi-refringent polar-ization-maintaining fibers," *Applied Optics*, Vol. 32, 1993, pp. 2213–2218.

[9] Schott Optical Glass, Inc. Catalog.

[10] Pozzi, F., C. De Bernadi, and S. Morasca. "New simple technique to measure group effective indices in optical fibers," *Electronics Letters*, Vol. 28, 1992, pp. 1947–1949.

[11] Heffner, B. L. "Attosecond-resolution measurement of polarization mode dispersion in short sections of optical fiber," *Optics Letters*, Vol. 18, 1993, pp. 2102–2104.

[12] Shibata, N., M. Tateda, and S. Seikai. "Polarization mode dispersion measurement in elliptical core single mode fibers by a spatial technique," *IEEE Journal Quantum Electronics*, Vol. QE-18, 1982, pp. 53–58.

[13] Kaminow, I. P. "Polarization in optical fibers," *IEEE Journal Quantum Electronics*, Vol. QE-17, 1981, p. 15.

[14] Cowap, S. F., and S. D. Brown. "Technique for the static fatigue testing of fibers," *Journal American Ceramic Society*, Vol. 70, 1987, pp. C67–C68.

[15] Alavie, A. T., and B. G. Grossman. "Accurate measurement of effective cutoff in elliptical core fibers," *Optics Letters*, Vol. 18, 1993, pp. 343–345.

[16] Dyott, R. B. "Method of determining the azimuthal position of the transverse axes of optical fibers with elliptical cores," U. S. Patent 5323225, June 21, 1994.

Chapter 7
The D Fiber

7.1 THE D FIBER

Two important requirements for polarization-preserving fibers are the accurate location of the birefringent axes and some method of coupling fibers laterally to form power splitters or directional couplers, which also hold the direction of polarization.

A way of meeting both goals is to make a fiber with an accessible guiding region and with a cross section that automatically locates the birefringent axes [1]. One such fiber is shown in section in Figure 7.1. The section is D-shaped with the flat of the D parallel to one of the axes of the elliptical core. The major axis is generally chosen because of the slightly tighter guidance of the $_oHE_{11}$ mode, making it the chosen mode of operation. The axes of birefringence are located by positioning the D fiber with the flat of the D against some surface. In order to get access to the guiding region, the fiber is made with the flat close to the cladding so that removing a small amount of material from the periphery of the section (for instance, by etching) exposes the evanescent fields.

7.2 FIBER MANUFACTURE

A preform for a standard elliptically cored fiber has a flat ground parallel to the major axis of the elliptical core. A preliminary coarse grind to remove the bulk of the material is followed by a fine-grain finishing cut to avoid producing variations in the fiber dimensions that could cause either scattering or coupling between modes, leading to the loss of polarization holding.

Subsequently, the ground preform is aligned carefully in the fiber-drawing chuck so that the flat of the drawn fiber lies on the surface of the lower wheel. Generally, the fiber is dual-coated with plastic, in which case additional location of the D flat (provided by a system of wheels and belts) is necessary just before the fiber enters the first coating die. This system avoids the periodic twisting of the fiber during drawing that causes problems

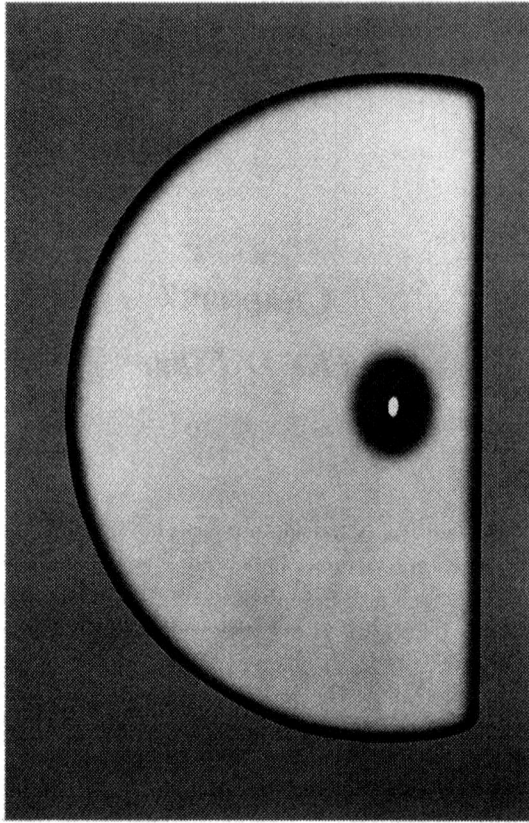

Figure 7.1 Photograph of a section of D fiber.

with the manufacture of components such as directional couplers. A length of drawn fiber is seen to be slightly bowed when suspended from one end. The effect is caused by the mismatch in thermal expansion between the silica body and the doped guiding region, which is displaced from the neutral axis of the D section. The fiber is tested for twist by suspending a length as shown in Figure 7.2. The resultant ''horseshoe'' should lie in one plane.

7.3 MEASUREMENTS

Measurements of attenuation, higher mode cutoff, birefringence, and polarization holding are made as for standard round section elliptically cored fiber (designated ''E'' fiber). Additionally, a test is made for the alignment of the flat of the D with the axes of birefringence. Figure 7.3 shows the arrangement.

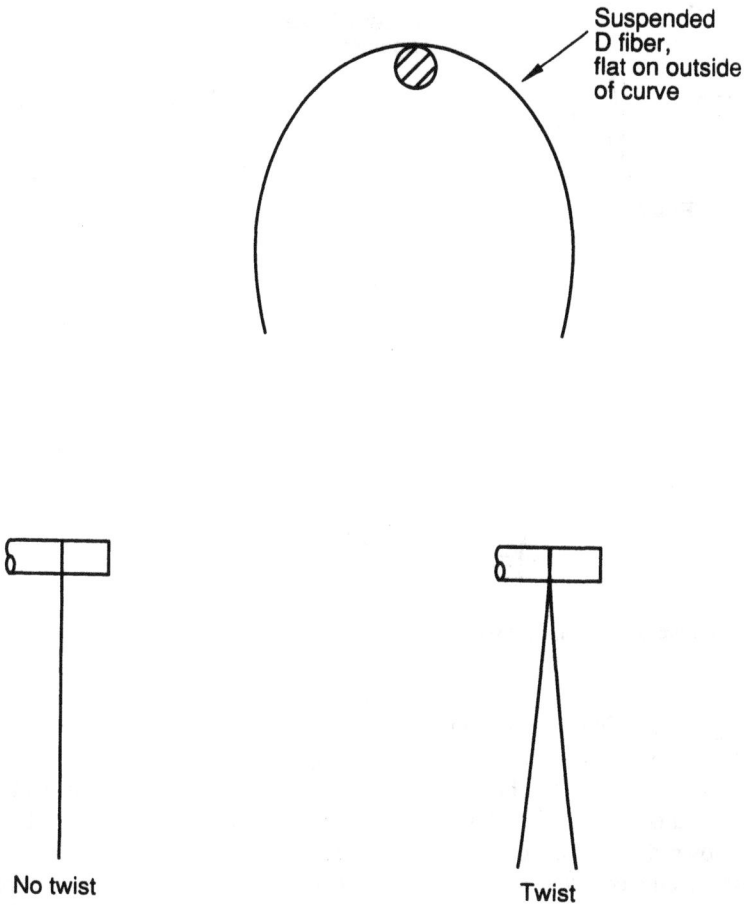

Figure 7.2 Method of testing for drawing-induced twist.

The fiber is placed with the flats of the D against the launching plates and the input and output polarizers are adjusted for minimum signal. The fiber ends are then interchanged and the process repeated. The change in angle, $\Delta\theta$, of the polarizers is noted. The angle between the flat and birefringent axis is then $\Delta\theta/2$.

7.4 ETCHING THE FIBER

As drawn, the fiber has its guiding region totally enclosed within the section, with a small thickness of silica between the fluorine-doped silica cladding and the flat surface of the D. It can be etched in a weak (10%) solution of hydrofluoric acid in water in order to expose

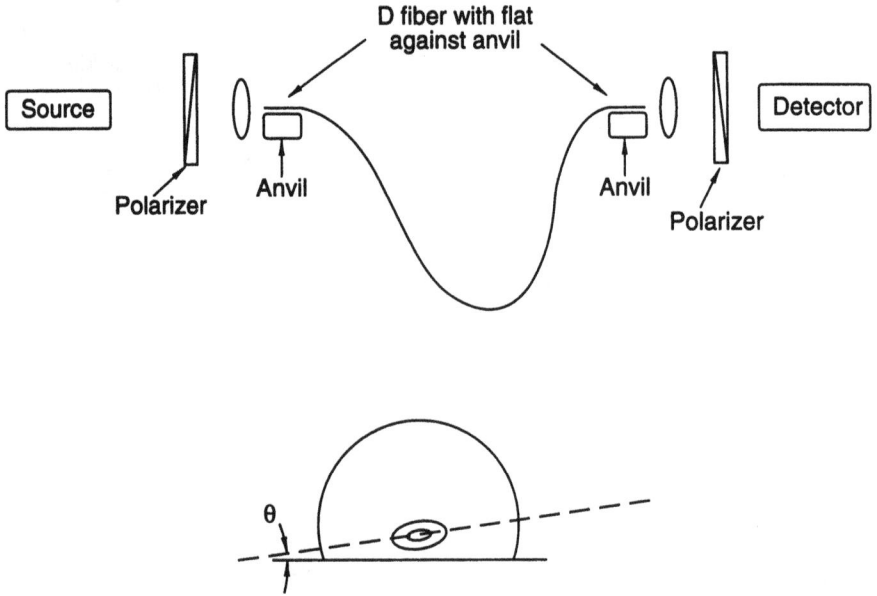

Figure 7.3 Measurement of angle between birefringent axis and the flat of the D section.

the guiding region. The acid bath must be kept at a constant temperature throughout the etch, as the rate of etch increases rapidly with temperature. The etching time may be found by trial and error and must be repeatable to within a few seconds in a thousand for consistent results. Alternatively, the etch can be monitored optically by the following method, shown diagrammatically in Figure 7.4.

Light from a coherent source is launched at 45° to the birefringent axes of the fiber to be etched, exciting both fundamental modes equally. The source wavelength is chosen to produce a strong evanescent field in the fiber; that is, the V value should be at the lower end of its operating range. The source should be as coherent as possible so that the decoherence length L_D is much longer than the fiber length L, where

$$L_D = \frac{L_S}{\Delta\overline{\beta} + V\dfrac{\partial(\Delta\overline{\beta})}{\partial V}} \tag{7.1}$$

where L_S = coherence length of the source and $\Delta\overline{\beta}$ = normalized difference in propagation constant of the two fundamental modes; that is, the difference in effective index.

Light emerging from the other end of the fiber passes through a Soleil-Babinet

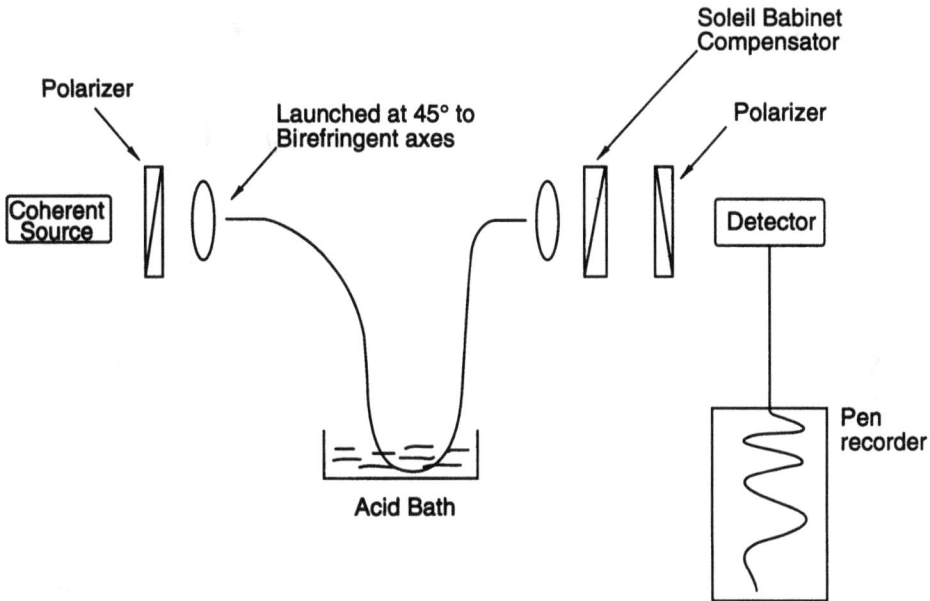

Figure 7.4 Method of monitoring fiber etching.

compensator and another polarizer onto a detector system connected to a chart recorder. A modulated source/phase-sensitive detection system can be used to improve sensitivity and to eliminate the effects of stray light.

The compensator and output polarizer are adjusted to give either a minimum or a maximum signal from the detector. The length of fiber to be etched is now immersed in the etchant. During the etch, first the silica boundary layer and then the cladding of the fiber guiding region (having a refractive index of about 1.5) are replaced by the acid/water solution with an index of about 1.33. When the acid reaches the evanescent field region, the phase velocity of both fundamental modes is increased. However, the even $_e\text{HE}_{11}$ mode has an evanescent field extending further than that of the odd $_o\text{HE}_{11}$ mode so that a phase shift is introduced between the modes causing a rotation of the polarization of the light emerging from the fiber, which is converted by the output polarizer into a change in amplitude of the detected signal.

Figure 7.5 shows a typical plot of detector output versus time. The periodic variations become more and more closely spaced until eventually the acid reaches the core and the signal disappears.

This method does not eliminate the need to calibrate the etch, but it does take out such variables as temperature and acid strength and is very repeatable. Several fibers can be etched simultaneously using just one fiber as the etch monitor.

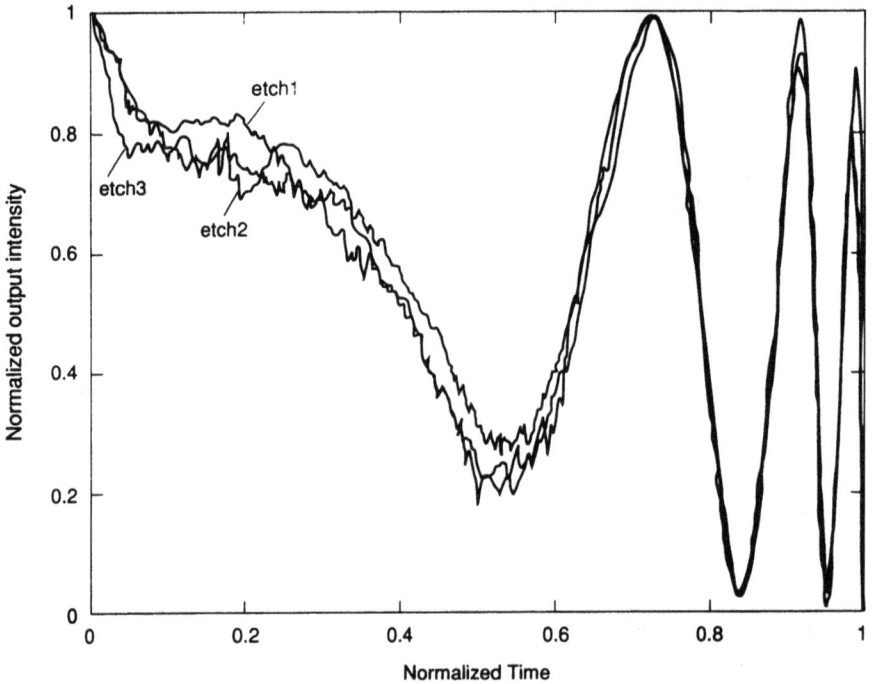

Figure 7.5 Plot of etch monitor.

7.5 COUPLING BETWEEN FIBERS

Polarization-holding fibers such as the elliptically cored fiber are used extensively in interferometers where it is essential to provide a single path in both arms in order to maintain a stable phase relationship between them. In other words, a polarization-holding fiber is not much use without a polarization-holding directional coupler.

A general theoretical study of coupling between parallel waveguides was made by Miller in 1954 [2]. Prior to that the coupling between two surface wave lines, similar in many ways to that between fibers, was analyzed by Meyerhoff in 1952 [3]. For weak coupling, where the propagation constant of the individual transmission line is not affected by the presence of its partner, there is a periodic and complete interchange of energy between lines that have equal propagation constants. The application to the coupling between monomode optical fibers with circular cores was done by Jones [4] in 1965 with some corrections by Burke [5] in 1967. Burke's results, taken from [5], are shown in Figure 7.6. They are based on the Jones analysis and also the prior analysis by Bracey et al. [6] on the coupling between parallel dielectric rods. The values in Figure 7.6, which are above and to the right of the dotted line, are accurate to within about 10% (the

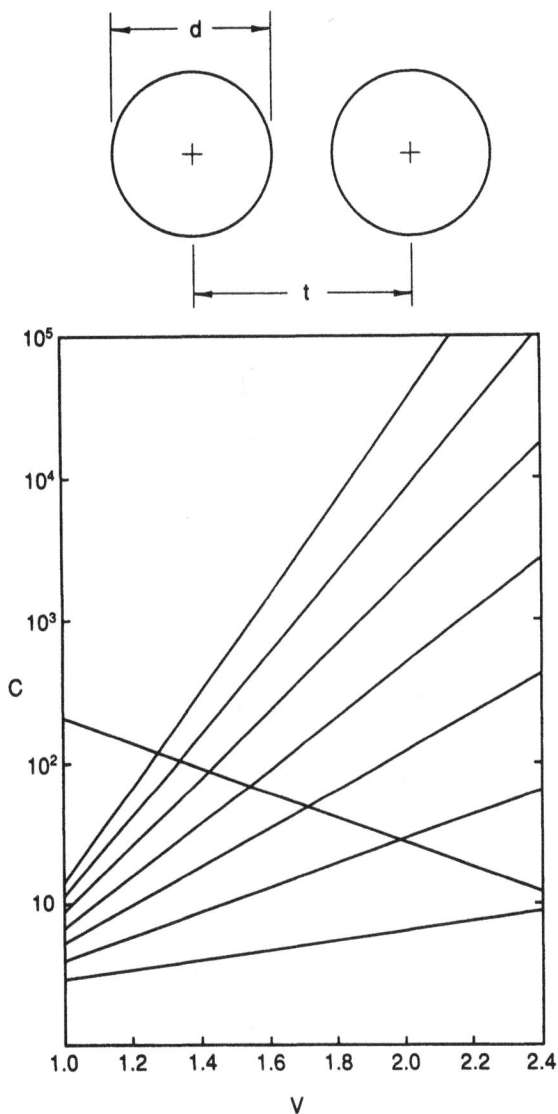

Figure 7.6 Theoretical coupling between circular core fibers [5].

accuracy being limited by an approximation in the first term of an asymptotic expansion of the Bessel K function, which is present in both [4] and [5]). Above this line, the beat length L_C (i.e., the length along the fibers for one complete cycle of interchange of energy out of one fiber into the other and back again) is given by

$$L_C = Ce^{\alpha V} \cdot \frac{\lambda_0}{n \Delta n} \tag{7.2}$$

where, as usual

$$V = \frac{2\pi a}{\lambda_0}[n_1^2 - n_2^2]^{\frac{1}{2}} \tag{7.3}$$

and

$$\alpha = 0.74 + 2.3\left[\left(\frac{t}{2a}\right) - 1\right] \tag{7.4}$$

where t is the center-to-center separation of the fiber cores, and a is the core radius. For example, for fibers with $n = 1.48$, $\Delta n = 0.03$ with cores spaced 1.0 diameter apart,

$$t/d = 1 \text{ at } V = 2.4 \ \lambda_0 = 850 \text{ nm}$$

Then

$$C = 8.2$$

$$\alpha = 0.74$$

$$L_C = 0.9 \text{ mm}$$

with $t/d = 1.5$; $V = 2.0$

$$C = 25$$

$$\alpha = 1.89$$

$$L_C = 2.1 \text{ cm}$$

The coupling length is most sensitive to the inter-core spacing. There seems to be no published method of analyzing the coupling between fibers with elliptical cores. An approximate estimate of coupling could be made by interpolation between the results for the circular core and infinite slab. Snyder [7] has published an extensive general analysis on coupled mode theory for optical fibers. A comprehensive selection of papers is available in the *SPIE Milestone Series* [8].

7.6 THE D-FIBER DIRECTIONAL COUPLER

A practical fiber directional coupler for use in interferometers should have a coupling coefficient that can be adjusted accurately during manufacture and then fixed permanently with no appreciable variation over a wide range of temperatures or with other external influences such as the twisting or bending of the fiber legs. It should hold polarization to a degree comparable with the fiber itself and should have a loss that is less than 0.3 dB. It should also be compact and inexpensive to make. In order for a coupler to meet these requirements, the fibers must be brought together so that their evanescent fields overlap and they must be held in position under all conditions. Unlike the usual monomode tapered fibers where the original guiding structure becomes so small that guidance is transferred to the pulled-down section, which acts as a core with the surrounding material (generally free space as the cladding), two D fibers positioned flat to flat can make a true evanescent field coupler; that is, one where the propagation characteristics of the two paths are only slightly changed within the coupling region (Dyott and Bello [9]).

Fused silica has such a small coefficient of thermal expansion ($0.5 \times 10^{-6}/°C$) that it is almost impossible to find materials with expansion coefficients that match over a working temperature range of, say, -55 to $+85°C$. This precludes glues and epoxies and constraining tubes of low-melting-point glass. Even if such materials do create a bond that survives over the temperature range, the resultant forces exerted on the fiber severely affect the polarization holding characteristics of the fibers and therefore the coupler.

7.6.1 Fusion Coupler

The problems of temperature stability and thermal mismatch can be overcome by positioning two D fibers flat to flat and then fusing them together with, say, a small flame or an electric arc. Such couplers are very stable thermally but suffer from variable coupling due to change in refractive index in the coupling region caused by stress from bending or twisting the fibers. Generally, some form of support is necessary, such as a "splint" in the form of a surrounding tube or channel. Figure 7.7 shows one such arrangement with the fibers positioned in a tube of Vycor glass, which has a thermal expansion coefficient almost exactly matching that of the silica fibers but with a softening point some 100°C below that of silica. The tube is heated with a vacuum applied to collapse and fuse it onto the fibers and so hold them together (Dyott, Handerek, and Bello [10]).

7.6.2 Coupler Tuning

Once the fibers have been brought together, some means must be used to adjust the coupling coefficient to give the needed power splitting. To do this, it is necessary to expand the evanescent fields to increase their overlap and hence the amount of coupling.

One way of doing this is by heating and pulling to reduce the diameter of the fibers

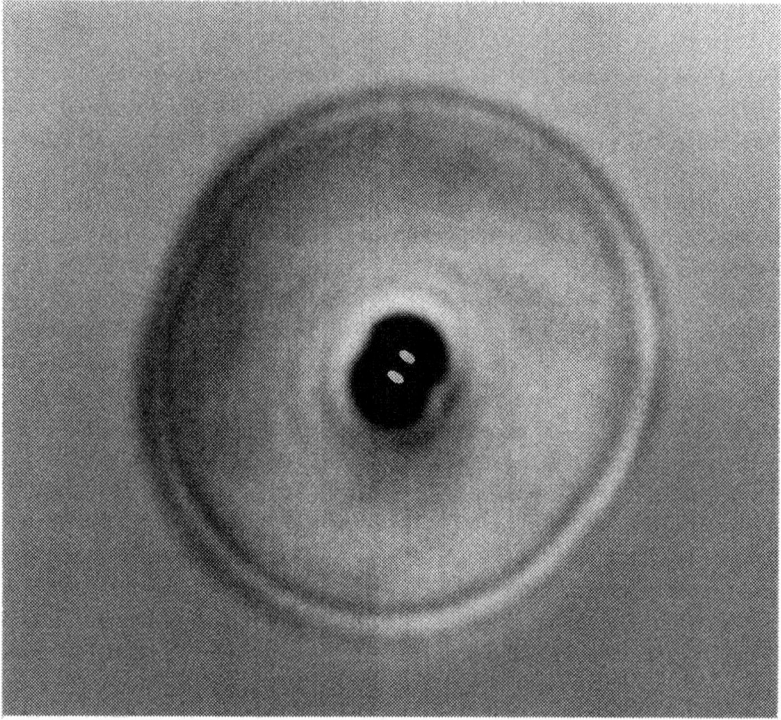

Figure 7.7 Photo of cross section of fused coupler.

and so extend the evanescent field [10]. Figures 4.12 and 4.13 show the power confinement $\eta = \dfrac{P_{core}}{P_{total}}$ for the odd and even fundamental modes as a function of V for different ellipticities demonstrating the rapid expansion of the evanescent field in the cladding as V, directly proportional to overall fiber dimensions, is reduced.

Another method of extending the evanescent field is to heat the coupler ensemble to a temperature where the germania-doped core diffuses into the cladding (Handerek and Dyott [11]).

The following analysis is for circular geometry (Chan et al. [12]), but will give some idea of the effect in elliptical geometry.

If the diffusion coefficient D is independent of germania concentration and the initial concentration is N_0, then the concentration N at radius r and time t is given by

$$\frac{\partial^2 N}{\partial r^2} + \frac{1}{r}\frac{\partial N}{\partial r} - \frac{1}{D}\frac{\partial N}{\partial t} = 0 \qquad (7.5)$$

Assuming that the cladding is infinite and that the following boundary conditions apply when $t = 0$,

$$N = N_0 \text{ for } 0 \leq r \leq a$$

$$N = 0 \text{ for } r > a$$

then the solution to (7.1) (derived from a solution by Carslaw and Jaeger [13] of a related heat transfer problem) is

$$\frac{N}{N_0} = \int_0^\infty \exp\left(-\frac{Dt}{a^2}u^2\right) J_0\left(\frac{ur}{a}\right) J_1(u) \, du \tag{7.6}$$

Assuming that the refractive index of the core depends linearly on the concentration of dopant, the index n is given by

$$n = (n_1 - n_2)\frac{N}{N_0} + n_2 \tag{7.7}$$

Figure 7.8 shows the relative concentration as a function of normalized radius r/a for different values of $\frac{Dt}{a^2}$. At $\frac{Dt}{a^2} = 0.1$, there is an approximately parabolic variation of refractive index with radius (Dyott and Brain [14]).

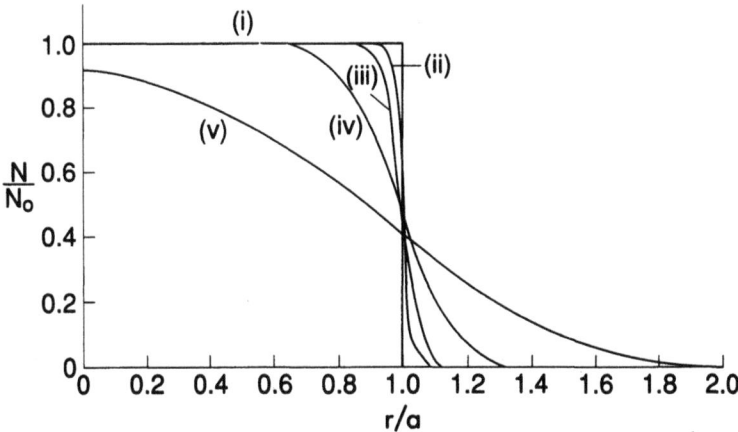

Figure 7.8 Diffusion in optical fibers. Curve(i), $Dt/a^2 = 0$; curve(ii), $Dt/a^2 = 10^{-4}$; curve(iii), $Dt/a^2 = 10^{-3}$; curve(iv), $Dt/a^2 = 10^{-2}$; curve(v), $Dt/a^2 = 10^{-1}$. (From [12].)

The effect of diffusion on the propagation constant of the fiber is negligible for $\frac{Dt}{a^2} < 10^{-3}$. Figure 7.9 taken from [12] shows the variation of $\bar{\beta}$ with diffusion for a fiber $n_1 = 1.53$, $n_2 = 1.50$.

7.6.3 Monitoring the Coupling

The coupling can be monitored by the straightforward method shown in Figure 7.10. Light at the operating wavelength of the coupler is launched into one coupler leg and the output monitored on the two legs on the other side of the coupler. If, for instance, the coupler is being tuned by diffusion, heat (usually from a flame) is applied to diffuse and then removed in order to measure the coupler at room temperature. Figure 7.11 shows a typical chart.

A more sophisticated method uses an optical spectrum analyzer in which case the variation of coupling over a wide range of wavelengths can be seen while the coupler is being tuned. Using this method, it is possible to make wavelength-division-multiplex (WDM) couplers with coupling at one wavelength and not at another.

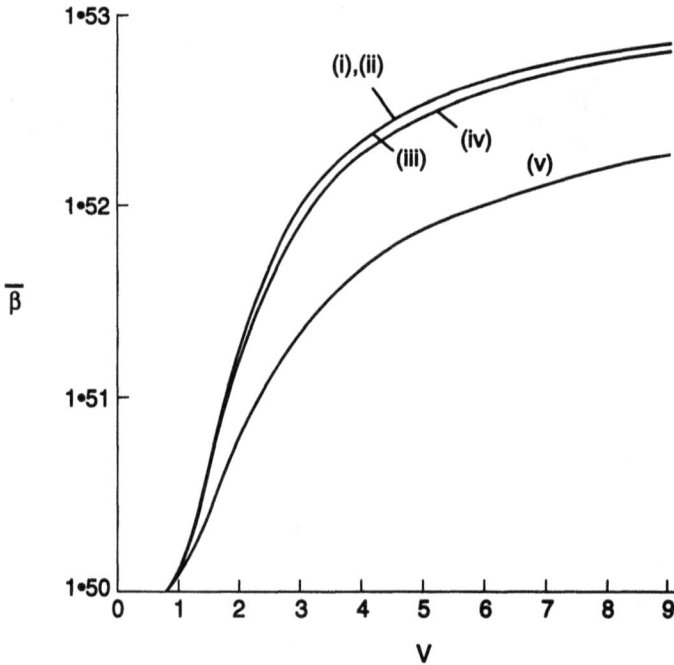

Figure 7.9 Effect of diffusion on normalized propagation constants. Curve(i), $Dt/a^2 = 0$; curve(ii), $Dt/a^2 = 10^{-4}$; curve(iii), $Dt/a^2 = 10^{-3}$; curve(iv), $Dt/a^2 = 10^{-2}$; curve(v), $Dt/a^2 = 10^{-1}$. (From [12].)

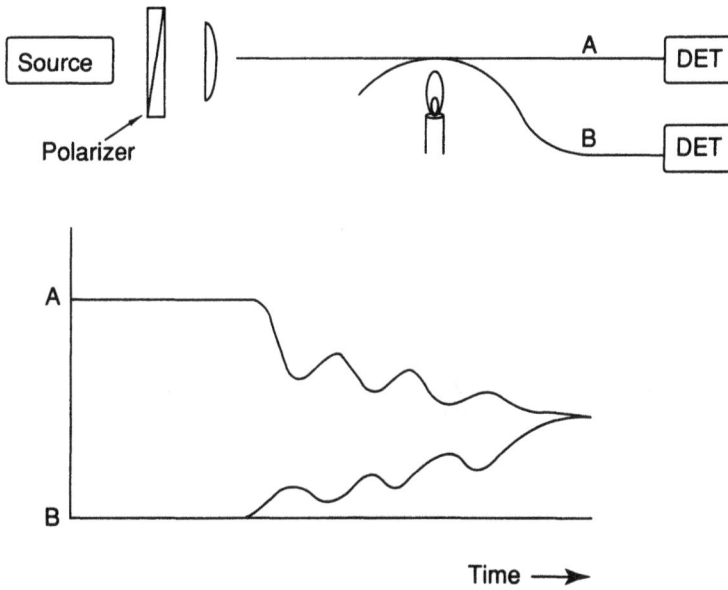

Figure 7.10 Method of the diffusion tuning of a coupler.

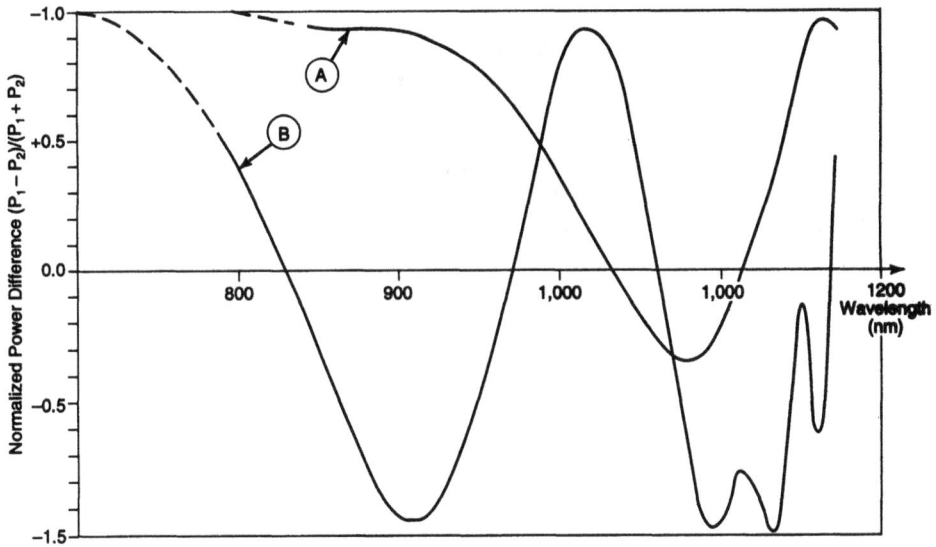

Figure 7.11 Spectral response of coupler: (**A**) Before diffusion, (**B**) after diffusion.

If a coupler with an equal division of power is needed, then the principle of the loop mirror can be used.

7.7 LOOP MIRROR

The loop mirror [10, 15] has many applications apart from monitoring coupling coefficients, and the D fiber is particularly suitable for its fabrication. Figure 7.12 shows the arrangement. A fiber is looped back on itself to form a coupler on the loop. Bearing in mind that a wave passing across the coupler has a phase delay of $\pi/2$ relative to a wave passing through but not across, and considering the clockwise and counterclockwise paths around the loop, it is easy to see that (in Figure 7.12), the counterclockwise path to coupler leg B is delayed by $\pi/2 + \pi/2 = \pi$ relative to the clockwise path so that there is zero output at B when the split in power is equal.

Assuming a lossless system, all the incident power must return along fiber A, thus forming a mirror. A proof of the $-\pi/2$ phase shift is given by Miller [2], but there is a simple heuristic argument, presented as follows.

Consider two coupled systems, A and B, which can be fibers or even two linked pendulums. Assume that energy is transferring from A to B and consider the following possible phase relationships between A and B.

1. *A and B in phase.* Clearly this cannot be so because it is then impossible to tell whether energy is passing from A to B or vice versa.
2. *A and B in antiphase.* The system is again symmetrical and the same argument applies.
3. *A and B in quadrature.* This is the next possibility. The system is asymmetric. Then A must lead B by $\pi/2$ because the energy is passing from A to B. If B initially has no energy, it cannot pass any to A.

The principal of energy transfer is illustrated by the dictum "The grandfather clock stops on Thursday." In Victorian households, Father winds the clock on Sundays. By Thursday, the weights have descended to a point level with the pendulum bob. A slight

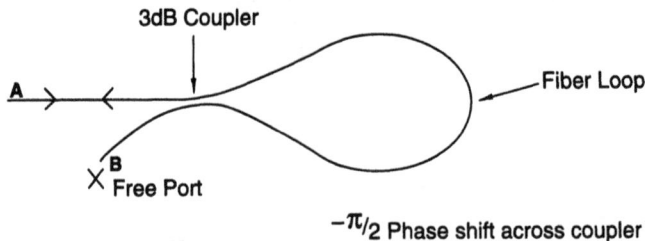

Figure 7.12 Loop mirror.

give in the clock case then couples the two systems and the energy in the pendulum transfers over to swing the weights. Transferring the energy back to the pendulum loses enough swing for the pendulum not to release the escapement and so the clock stops.

7.8 INDIUM-COATED D-FIBER POLARIZER

It is possible to make a simple but effective fiber polarizer by coating an etched D fiber with indium to form a plasmon guide where the mode producing an electric field normal to the fiber-metal surface is suppressed, with the orthogonal mode being transmitted with low attenuation (Dyott, Bello, and Handerek [16]).

7.8.1 Properties of Indium

The optical properties of indium are listed for a polished crystal (Lenham and Treherne [17]) and for a deposited thin film (Motulevich [18]). The refractive index of a metal is complex $n + j\chi$.

In Table 7.1, the refractive index of indium interpolated from [17, 18] is compared with values given by Born and Wolf [19] for other metals at a wavelength of 590 nm. Table 7.2 from de Brunijn et al. [20] gives the values of n and χ for indium at room temperature over a wavelength range of 420 to 1,560 nm. The indium (melting point 157°C) is applied to the fiber by passing it through a molten pool of the metal. The indium adheres to the fiber by means of a thin film of indium suboxide. No values for the refractive index of indium suboxide could be found by the author.

Figure 7.13 shows a diagram of the polarizer where the flat of the D is parallel with the major axis of the elliptical core. The $_eHE_{11}$ mode is suppressed and the $_oHE_{11}$ (or slow mode) is passed. Results from a polarizer 70-mm long give a measured polarization

Table 7.1
Refractive Index of Various Metals at 590 nm

Metal	n	χ
Indium*	0.795	5.02
Silver[†]	0.795	3.44
Aluminum[†]	0.20	5.23
Tin[†]	1.44	5.25
Gold[†]	0.47	2.83
Cadmium[†]	1.13	5.01
Gallium[†]	3.69	5.43 (single crystal)

*From [17] and [18].
[†]From [19].

Table 7.2
Refractive Index of Indium at Various Wavelengths

λnm	n	χ
420	0.65	1.84
600	1.29	2.59
710	1.38	6.24
1050	1.83	7.94
1560	2.31	11.3

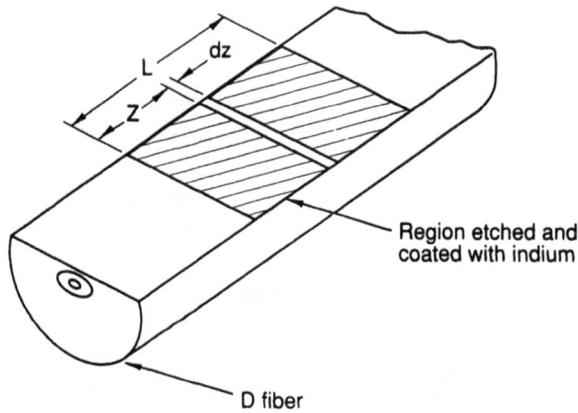

Figure 7.13 Indium-coated D-fiber polarizer.

suppression of about 40 dB with an insertion loss for the wanted mode of about 1.3 dB. However, the actual suppression of the incoming $_eHE_{11}$ mode may be far greater because the measured value is limited by the coupling from the unsuppressed $_oHE_{11}$ mode to the $_eHE_{11}$ mode in the polarizer itself. The following is an analysis of the situation.

7.8.2 Mode Coupling Within the Polarizer

The assumptions made are that the coupling is incoherent and that it is constant along the length L of the polarizer. The criterion for the first assumption is that the coherence length of the source should be much shorter than the length of the polarizer $(\lambda_0^2/\Delta\lambda) \ll L$. This would be true, for instance, for an SLD operating at $\lambda_0 = 820$ nm with a line width $\Delta\lambda = 20$ nm, giving a coherence length of about 33 μm and a polarizer length of 70 mm in a typical gyro application. The second assumption is valid if the intermode coupling is constant over the length of the polarizer, which should be true if the indium coating is applied evenly. Let

P_e = power in the $_e\text{HE}_{11}$ mode

P_o = power in the $_o\text{HE}_{11}$ mode

P_i = input power launched into each mode equally at $z = 0$

α_e = attenuation in nepers/m of the $_e\text{HE}_{11}$ modes

Then the amount of mode power P_e reaching a distance z meters to an elemental length dz is

$$P_e(z) = P_i \exp(-\alpha_e z) \tag{7.8}$$

Let α_t = attenuation in nepers/m of the $_o\text{HE}_{11}$ mode due to mode conversion to $_e\text{HE}_{11}$.

Assume that the polarizer is lossless except for this transferred power. Then the amount of power P_o reaching dz is

$$P_o(z) = P_i \exp(-\alpha_t z) \tag{7.9}$$

The amount of power P_o transferred to P_e in element dz is

$$\Delta P_e(z) = P_i \exp(-\alpha_t z)[1 - \exp(-\alpha_e dz)] \tag{7.10}$$

This power (in the even mode) is attenuated by the polarizer so that the amount reaching L is

$$\Delta P_e(L) = P_i \exp(-\alpha_t z - \alpha_e(L - z))[1 - \exp(-\alpha_t dz)] \tag{7.11}$$

For small dz,

$$\exp(-\alpha_t dz) = 1 - \alpha_t dz \tag{7.12}$$

Then

$$\Delta P_e(L) = \alpha_t P_i \exp(-\alpha_e L + [\alpha_e - \alpha_t]z)dz \tag{7.13}$$

Integrating the transferred power from each element over length L, the total emerging transferred power P_t is

$$P_t = \alpha_t P_i \exp(-\alpha_e L) \int_0^L \exp(\alpha_e - \alpha_t) z\, dz \tag{7.14}$$

$$P_t = \left[\frac{\alpha_t}{\alpha_e - \alpha_t}\right] P_i \exp(-\alpha_e L) \left[\exp\left(\left[\alpha_e - \alpha_t\right]L\right) - 1\right] \tag{7.15}$$

Also, the power in the $_e\text{HE}_{11}$ mode reaching L from $z = 0$ is

$$P_e(L) = P_i \exp(-\alpha_e L) \tag{7.16}$$

when $\alpha_e \gg \alpha_t$. Then the total even mode power P_E at L is

$$P_E = P_e(L) + P_T \tag{7.17}$$

Since $\alpha_e \gg \alpha_t$,

$$P_E \simeq P_i \exp(-\alpha_e L) + \left(\frac{\alpha_t}{\alpha_e}\right) P_i [1 - \exp(-\alpha_e L)] \tag{7.18}$$

The power in the odd mode reaching L is

$$P_o(L) = P_i \exp(-\alpha_t L) \tag{7.19}$$

Again, neglecting α_t compared with α_e,

$$\frac{P_e}{P_o(L)} = \exp(-\alpha_e L) + \left(\frac{\alpha_t}{\alpha_e}\right) \exp(\alpha_t L) \left[1 - \exp(-\alpha_e L)\right] \tag{7.20}$$

The polarization suppression ratio S is conventionally expressed as the inverse of this ratio:

$$S = \frac{P_o(L)}{P_e} \tag{7.21}$$

Figure 7.14 shows a plot of S_{dB} against length L of the polarizer for various values of α_e and α_t. The initial slope of the curves gives the rate of attenuation α_e of an incoming $_e\text{HE}_{11}$ mode.

The critical length of the polarizer at the knee of the curve where saturation occurs is given by

$$\exp(-\alpha_e L) = \left(\frac{\alpha_t}{\alpha_e}\right) \exp(\alpha_t L)[1 - \exp(-\alpha_e L)] \tag{7.22}$$

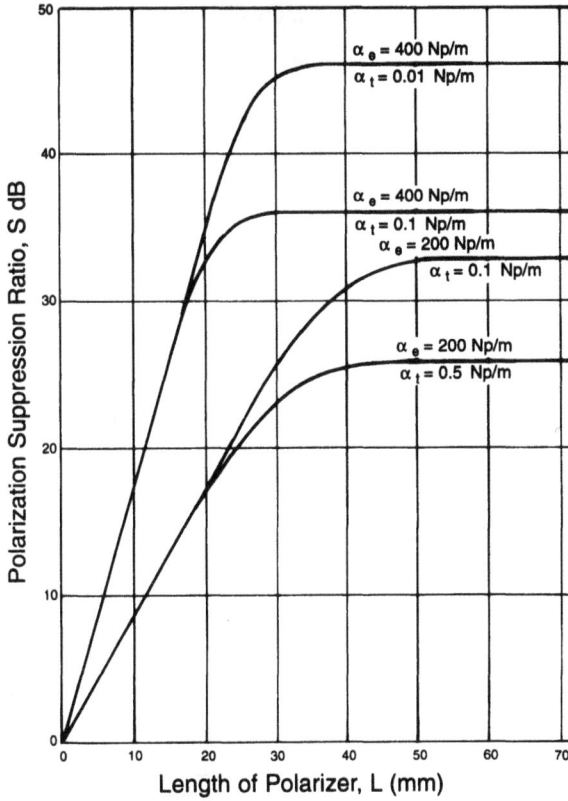

Figure 7.14 Theoretical variation of suppression ratio with polarizer length for various values of α_e and α_t.

Since for large $\alpha_e L$

$$1 - \exp(-\alpha_e L) \simeq 1 \qquad (7.23)$$

then

$$\exp[(\alpha_e + \alpha_t)L] = \frac{\alpha_e}{\alpha_t} \qquad (7.24)$$

with $\alpha_e \gg \alpha_t$

$$L = \left(\frac{1}{\alpha_e}\right) \ln\left(\frac{\alpha_e}{\alpha_t}\right) \qquad (7.25)$$

and the saturated polarization suppression ratio is

$$S_{\text{sat}} \simeq \frac{\alpha_e}{\alpha_t} \tag{7.26}$$

Figure 7.15 shows an experimental plot of polarization suppression versus length for an indium-coated fiber polarizer cut back in steps. The solid line is the theoretical suppression taking $\alpha_e = 358$ nepers/m = 1556 dB/m and $\alpha_t = 0.036$ nepers/m = 0.155 dB/m.

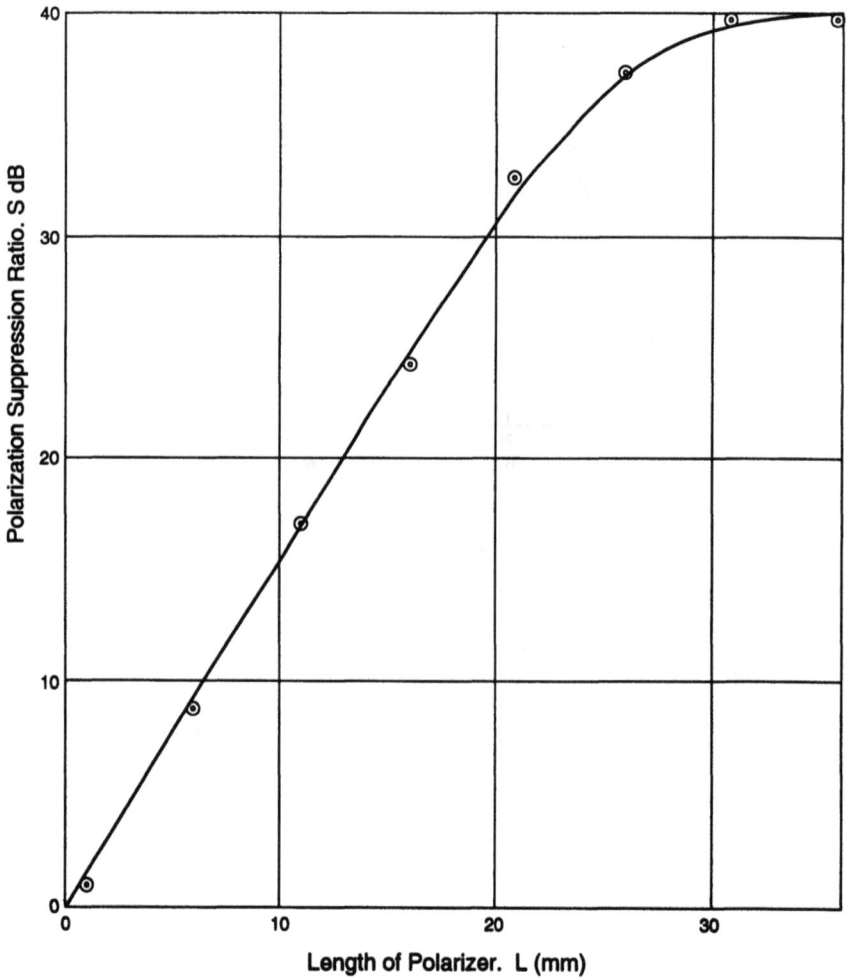

Figure 7.15 Measured suppression ratio of polarizer.

The polarizer has a dual action, continually suppressing the power of the incoming $_e\mathrm{HE}_{11}$ mode and at the same time coupling power from the $_o\mathrm{HE}_{11}$ mode to the $_e\mathrm{HE}_{11}$ mode along the polarizer itself. This has interesting implications for the fiber gyro, where the purpose of the polarizer is to remove any power coupled from the $_o\mathrm{HE}_{11}$ mode to the $_e\mathrm{HE}_{11}$ in the gyro coil. Because of this dual action, such power will continue to be attenuated along the polarizer. However, the power coupled locally into the $_e\mathrm{HE}_{11}$ mode is unimportant in the sense that it is synchronous with that in the $_o\mathrm{HE}_{11}$ mode and will therefore not cause multiple path problems.

7.9 DIFFRACTION GRATINGS ON D FIBER

Gratings can be fabricated directly on to the D flat using a holographic method [21]. The fiber is etched to expose the evanescent field and photoresist is deposited by a pulling method described in [22, 23]. The coated fiber is then exposed to an interference pattern created as shown in Figure 7.16 from [21]. The exposed surface is developed and the diffraction grating so produced is used as a mask for chemical etching to give a more permanent grating, which also has the advantage of less absorption than does a grating from photoresist alone. After the etch, the photoresist is dissolved off the fiber. Figure 7.17 from [21] shows a SEM photograph of the etched grating. The pitch is 1.2 μm and

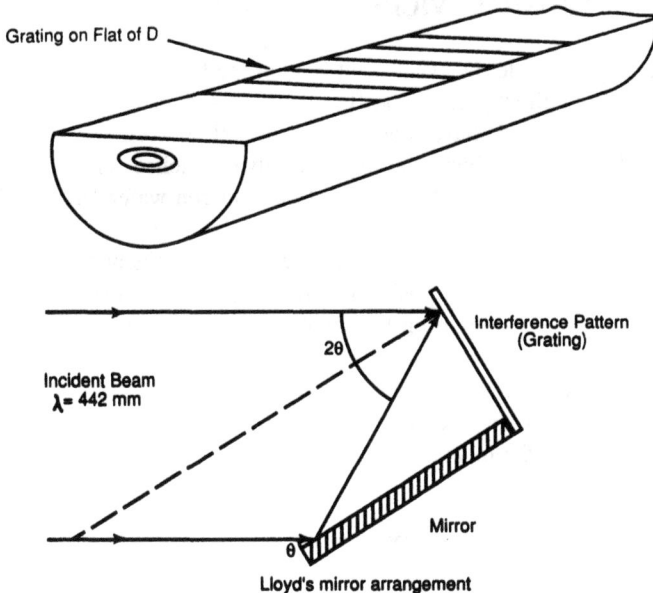

Figure 7.16 Diffraction grating on D fiber [21].

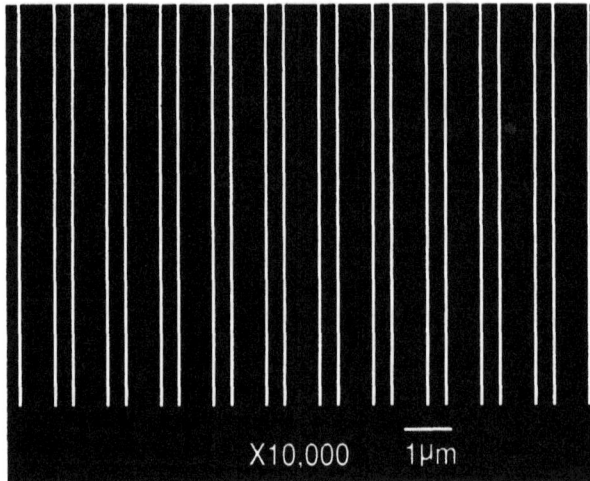

Figure 7.17 Photomicrograph of grating on D fiber [21].

the grating depth about 0.1 μm. An analysis of diffraction gratings on D fiber is given by Jensen and Selfridge [24].

7.10 OPTOELECTRONIC DEVICES ON D FIBER

Because of the easy access to the evanescent field of the D fiber, it is possible to construct optoelectronic devices directly on to the flat of the D, for instance, by the deposition of III-V structures. Li et al. [25] have reported interdigital photoconductive detectors using GaAs deposited by molecular beam epitaxy. The fiber is etched, as described previously, to expose the evanescent field and is mounted on a silicon wafer with the D flat uppermost.

About 2.5 μm of GaAs is deposited on the D flat, and is annealed to improve the material quality. The fiber is then positioned in a V groove on a silicon wafer as shown in Figure 7.18. Photoresist is spun on the whole wafer and standard lithography used to make the detector configuration as shown in Figure 3 of [25].

The in-fiber detector together with the gratings on the D bring integrated optics to the fiber itself, eliminating the need to break out of the fiber with the resultant problems due to reflections and to the mismatch between the geometries of the fiber and the chip.

7.10.1 Coupling of D Fiber to a Laser Diode

Because of its self-aligning properties, the D fiber is particularly suitable for direct butt coupling to the facet of a laser diode, thus eliminating the need for a focusing lens where

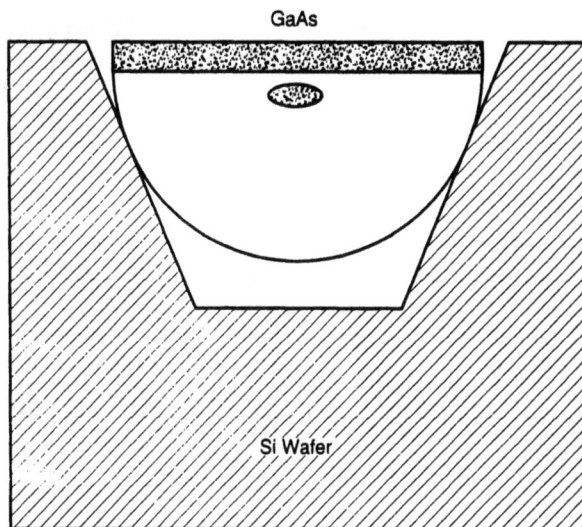

Figure 7.18 GaAs deposited on D fiber [25].

reflections can cause problems of laser stability due to optical feedback (as well as reducing the coupling efficiency). The flat of the D aligns the elliptical core with the elliptically shaped emission profile of the diode laser. The laser output is polarized in the plane of the junction so that coupling is to the $_o\text{HE}_{11}$ fiber mode, which has the transverse electric field along the major axis of the core.

Donhowe and Hunsperger [26] have studied the coupling from a laser diode to an elliptical fiber both theoretically and experimentally. Their analysis of the problem assumes a gaussian distribution of the field both for the fiber and for the laser using approximations by Sarkar, Pat, and Thyagarajan [27] and Sarkar, Thyagarajan, and Kumar [28] for the fiber and by Botez and Ettenberg [29] for the laser diode.

For a perfect butt coupling between fiber and laser, with no transverse or angular misalignment, the gaussian coupling efficiency η_g is [26]

$$\eta_g = \frac{2}{\left[\dfrac{w_{fx}}{w_{lx}} + \dfrac{w_{lx}}{w_{fx}}\right]} \cdot \frac{2}{\left[\dfrac{w_{fy}}{w_{ly}} + \dfrac{w_{ly}}{w_{fy}}\right]} \tag{7.27}$$

where w_{fx}, w_{lx} and w_{fy}, w_{ly} are the radii of the fiber (subscript f) and laser (subscript l) modal fields in the x and y directions, which can be derived from the far-field radiation patterns of fiber and laser.

Practical coupling efficiencies, defined as power out of the coupled fiber/power out of the uncoupled laser, of 65% have been achieved by gluing the end of the fiber to the laser facet using epoxy cement. Surprisingly, perhaps, such treatment does not seem to shorten the life of the laser, tested over several years.

7.10.2 Other Possibilities

Other possibilities for the D fiber include putting a nonlinear film on the etched D to produce either an optical switch or second harmonic generation, and the periodic deposition onto the D of a material with a high verdet constant at a pitch equal to the beat length between the fundamental modes. Applying a magnetic field along the fiber would then rotate the polarization via the Faraday effect for use as a current sensor or an in-fiber isolator, a much sought-after component (see Figure 7.19).

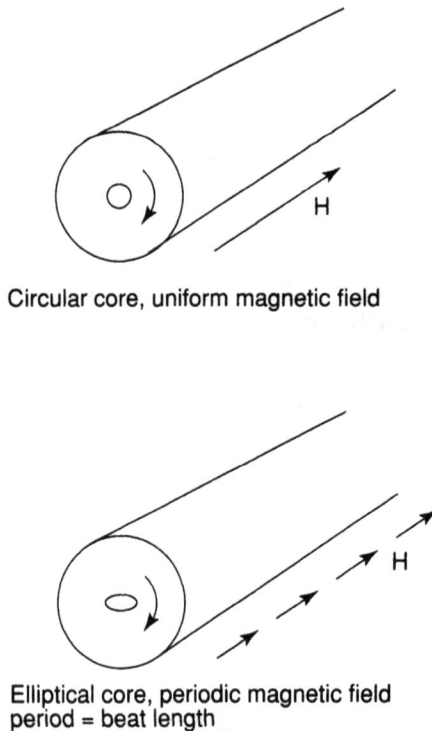

Circular core, uniform magnetic field

Elliptical core, periodic magnetic field
period = beat length

Figure 7.19 Rotation of polarization in elliptical fiber with periodic magnetic field.

REFERENCES

[1] Dyott, R. B., and P. F. Schrank. "Self-locating elliptically cored fiber with an accessible guiding region," *Electronics Letters*, Vol. 18, 1982, pp. 980–981.

[2] Miller, S. E. "Coupled wave theory and waveguide applications," *Bell System Technical Journal*, Vol. 33, 1954, pp. 661–719.

[3] Meyerhoff, A. A. "Interaction between surface wave transmission lines," *Proc. IRE*, Vol. 40, 1952, pp. 1061–1065.

[4] Jones, A. L. "Coupling and scattering in optical fibers," *Journal Optical Society of America*, Vol. 55, 1965, p 261.

[5] Burke, J. J. "Switching with fiber-optical waveguides," *Journal Optical Society of America*, Vol. 57, 1967, pp. 1056–1057.

[6] Bracey, J. F., A. L. Cullen, E.F.F. Gillespie, and J. C. Staniforth. "Surface-wave research in Sheffield," *IRE Transactions on Antennas and Propagation*, Vol. AP-7, 1959, pp. S219–225.

[7] Snyder, A. W. "Coupled-mode theory for optical fibers," *Journal Optical Society of America*, Vol. 62, 1972, pp. 1267–1277.

[8] "Selected papers on coupled mode theory in guided wave optics." *SPIE Milestone Series*, Vol. MS 84, Editor, B. J. Thompson.

[9] Dyott, R. B., and J. Bello. "Polarization holding directional coupler mode from elliptically cored fiber having a D section," *Electronics Letters*, Vol. 19, 1983, pp. 607–608.

[10] Dyott, R. B., V. H. Handerek, and J. Bello. "Polarization holding directional couplers using D fiber," *Proc. SPIE*, Vol. 479, 1984, pp. 23–27.

[11] Handerek, V. A., and R. B. Dyott. "Fused D fiber couplers," *Proc SPIE*, Vol. 574, 1985.

[12] Chan, K. B., P.J.B. Clarricoats, R. B. Dyott, G. R. Newns, and M. A. Savva. "Propagation characteristics of an optical waveguide with a diffused core boundary," *Electronics Letters*, Vol. 6, 1970, pp. 748–750.

[13] Carslaw, H. S., and J. C. Jaeger. *Conduction of Heat in Solids*, Oxford, 1947.

[14] Dyott, R. B., and M. C. Brain. "Refractive index profiles for diffused glass fibers," *Electronics Letters*, Vol. 10, 1974, pp. 131–132.

[15] Miller, I. D., D. B. Mortimer, P. Urquhart, J. Ainslie, S. P. Craig, C. A. Miller, and D. B. Payne. "A Nd^{3+} doped c.w. fiber laser using all-fiber reflectors," *Applied Optics*, Vol. 26, 1987, pp. 2197–2201.

[16] Dyott, R. B., J. Bello, and V. A. Handerek. "Indium-coated D-shaped fiber polarizer," *Optics Letters*, Vol. 12, 1987, pp. 287–289.

[17] Lenham, A. P., and D. M. Treherne. "The optical constants of aluminium and indium," *Proc. Phys. Soc.* Vol. 85, 1965, p. 167.

[18] Motulevich, G. P. "Optical properties of nontransition metals," *Proc. P.N. Lebedev Physics Institute.*, Vol. 55, 1973, p. 1.

[19] Born, M., and E. Wolf. *Principles of Optics*, Pergamon Press, 1970 Fourth Edition, p. 621.

[20] de Bruijn, H. E., R.P.H. Kooyman, and J. Greve. "Choice of metal and wavelength for surface, plasmon resonance sensors: some considerations," *Applied Optics*, Vol. 31, 1992, pp. 440–442.

[21] Freeze, J. D., and R. H. Selfridge. "D-fiber holographic diffraction gratings," *Optical Engineering*, Vol. 32, 1993, pp. 3267–3271.

[22] Freeze, J. D. "Fabrication of high efficiency diffraction gratings in single mode D fibers using a holographic beam interference approach," Master's Thesis, Dept. of Electrical and Computer Eng. Brigham Young University, 1991.

[23] Yang, C. C., J. Y. Josefowicz, and L. Alexandru. "Deposition of ultrathin films by a withdrawal method," *Thin Solid Films*, Vol. 74, 1980, pp. 117–127.

[24] Jensen, M. A., and R. H. Selfridge. "Analysis of diffraction gratings based on D shaped fiber," *Journal American Optical Society*, Vol. 9, 1992, pp. 1086–1090.

[25] Li, W. Q., A. Chin, P. Bhattacharya, and S. Divita. "Molecular beam epitaxial GaAs optical detectors on silica fibers," *Applied Physics Letters*, Vol. 52, 1988, pp. 1768–1770.

[26] Donhowe, M. N., and R. G. Hunsperger. "Coupling non-circular core optical fibers to laser diodes," *SPIE*, Vol. 988, 1988, pp. 201–208.

[27] Sarkar, S. H., B. P. Pal, and K. Thyagarajan. "Lens coupling of laser diodes to monomode elliptic core fibers," *Journal Optical Communications*, Vol. 7, 1986, pp. 92–96.

[28] Sarkar, S. H., K. Thyagarajan, and A. Kumar. "Gaussian approximation of the fundamental mode in single mode elliptic core fibers," *Optics Communications*, Vol. 49, 1984, pp. 178–183.

[29] Botez, D., and M. Ettenberg. "Beamwidth approximations for the fundamental mode in symmetric double-heterojunction lasers," *IEEE Journal Quantum Electronics*, Vol. QE-14, 1978, p. 827.

Chapter 8

Applications of Elliptical Core Fiber

8.1 APPLICATIONS OF ELLIPTICAL CORE FIBER

The uses of elliptical core fiber can be split into two categories: those where the fiber is a direct alternative to stress-induced birefringent fibers, and those that use the properties of the elliptical core per se. These categories can be split further into those that are extrinsic, where the fiber is used as a transmission line (say to preserve polarization) and those that are intrinsic, where the characteristics of the fiber itself are changed in some way. A comprehensive description of optical-fiber sensors is given in two volumes edited by Culshaw and Dakin [1].

An example where fibers are used in all four categories is the fiber-optic gyroscope (FOG).

8.2 FIBER-OPTIC GYROSCOPE

Many of the properties of elliptical core fiber are brought together in the FOG. The principle of the FOG, shown in Figure 8.1, is based on the Sagnac interferometer. A single path of light is split into two equal parts, which travel in opposite directions around a loop and recombine through the splitter into a single path again. The splitter-loop combination acts as a mirror (see Chapter 7) so that the light returns back down the path from where it came. If the loop is turned about an axis perpendicular to the plane of the loop, a phase difference $\Delta\varphi$ appears between the two paths:

$$\Delta\varphi = \frac{4\omega\Omega}{c^2}A \qquad (8.1)$$

where

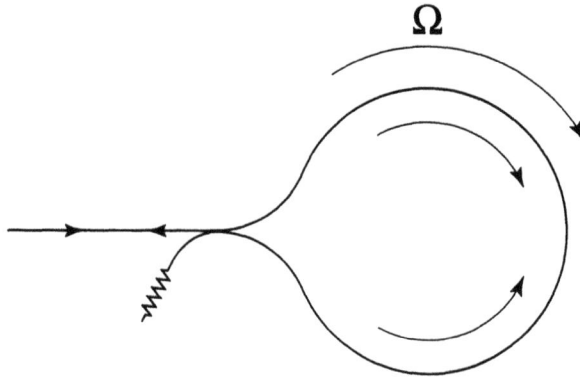

Figure 8.1 Sagnac interferometer in fiber.

ω = optical radian frequency, rad/sec

Ω = angular rotation rate, rad/sec

A = total area enclosed by the loop

c = velocity of light in free space

In the FOG, the loop is a coil of fiber. For a circular coil of N turns and diameter D_m, total path length L_m

$$A = N \frac{\pi D_m^2}{4}$$

$$N = \frac{L_m}{\pi D_m}$$

$$A = \frac{L_m D_m}{4}$$

$$\Delta\varphi = \frac{\omega\Omega}{c^2} L_m D_m \tag{8.2}$$

The phase shift is independent of the propagation constant of the medium through which the wave is traveling. A derivation of (8.2) for a medium other than free space, which involves a relativistic approach, is given by Lefèvre [2].

Rotating the loop about it's axis causes a phase shift between the clockwise and counterclockwise paths, and the loop no longer acts as a perfect reflector. The power of the reflected light diminishes by an amount that is transferred to the free arm of the splitter/coupler where, with no rotation, the output power is zero.

The reflected power is a function of the cosine of the phase difference so that the gyro used in the manner described is insensitive to small rates of rotation. In order to shift the point of operation from maximum power to maximum change of power with rotation rate and also to transform the system from one where the detection signal is dc to one where it is amplitude modulated (for greater convenience and stability), the phase of both paths around the loop is modulated at a frequency f_m and at an amplitude to produce a phase shift of $\frac{\pi}{2}$ to operate at maximum slope $\frac{dP}{d(\Delta\varphi)}$. The phase modulator is placed at one end of the loop or coil so that, due to the different times of transit through the modulator of the clockwise and counterclockwise guided waves, each receives a different phase modulation, thus creating the phase shift between them. A minimum amplitude of phase modulation is needed when the transit time for a wave traveling through the coil (at the group velocity) is equal to one-half cycle of the modulating frequency. Then, one wave receives a maximum positive phase modulation and its counterpart maximum negative modulation.

The minimum phase difference $\Delta\varphi$ detectable by a FOG is less than 10^{-6} rad, the limit being largely determined by shot noise in the detector. With wavelengths of a fraction of a micron and lengths of fiber of several hundred meters, the Sagnac interferometer works to phase differences between paths of one in about 10^{16}. It follows that anything that causes a phase shift of this order during the transit time of the light through the coil will affect one path differently from the other, causing phase shifts indicating spurious rotation rates and resulting in gyro "drift." Therefore, for there to be no phase shift not associated with rotation, there must be one path only through the interferometer in the sense that the phase delay of the path must not change by more than 10^{-6} rad during the transit time. There are two approaches to a single path. The first is the seemingly obvious one of choosing a single path among the several. The second is to scramble all the paths together so that there is just one average path.

A single-mode fiber with a perfectly circular core laid in a perfectly straight line has a single path. Bending the fiber in order to coil it causes stress, which in turn produces azimuthal changes of index in the guiding region that make the path length dependent on the polarization of the light. The solution, then, is to depolarize the light completely. Satisfactory gyros have been made to work using this principle although there are problems with changes of temperature and with magnetic fields affecting the vestiges of polarization that remain in the fiber. Change in temperature causes change in birefringence of the bent fiber, and a magnetic field along the fiber rotates the polarization via the Faraday effect.

The other approach is to preserve one polarization, and therefore one path, throughout the interferometer. This means using polarization-preserving fiber for the coil together with polarization-preserving components such as couplers and splices. Most important of all is a polarizer with an adequate suppression of the unwanted polarization. An analysis by Kintner [3] points out that polarization drift error in a gyro is proportional to the extinction ratio ϵ of the electric field rather than that of the optical power $E = \epsilon^2$, which is the measured extinction ratio. To an approximation, the magnitude of ϵ is equal to that of the minimum detectable phase shift in radians, which makes for the gyro limit $\epsilon \approx 10^{-6}$ and $E \approx 10^{-12}$, giving a power extinction ratio of -120 dB.

This degree of polarization suppression seems impossible to achieve and equally impossible to measure. However, returning to the indium-coated D-fiber polarizer from Chapter 7, Figure 8.2 (which is a repeat of Figure 7.15) shows how the measured extinction ratio of the polarizer reaches a saturated value along the length of the polarizer. This is because as the polarizer suppresses the unwanted polarization, it also regenerates it by random coupling from the wanted polarization within the polarizer itself. However, power that has been coupled into the unwanted polarization at any point before the polarizer (e.g., within the gyro coil and which will cause gyro drift) continues to be suppressed at the rate of the presaturation gradient of the polarizer; for example, if the initial gradient is 1,000 dB/m and the polarizer is 0.1m long, the actual suppression ratio for the gyro will be 100 dB rather than the measured saturated value of 40 dB.

At these high extinction ratios, it is important that the suppressed light does not

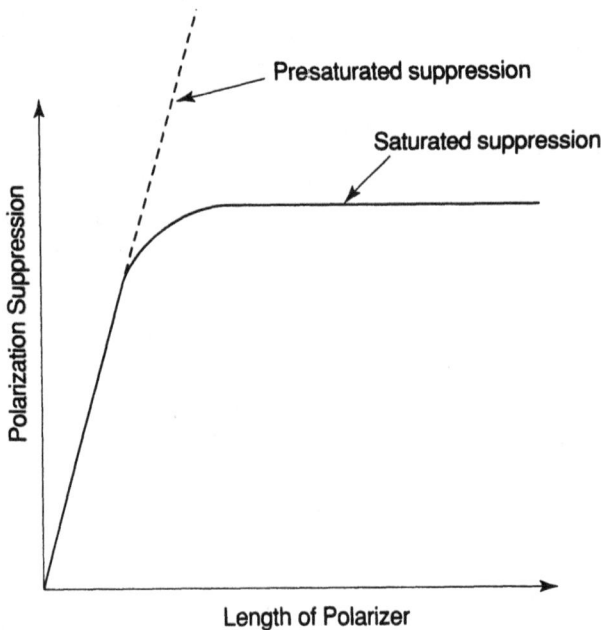

Figure 8.2 Saturation effects in polarizer.

somehow reenter the polarizer or the fiber beyond it, as may be so with polarizers that radiate the unwanted light energy rather than converting it into heat (as with the indium coated D). The extinction process is analogous to raking a sand pit. Each passage of the rake leaves the surface looking much the same generally, but obliterates the marks of the previous raking.

Figure 8.3 shows a diagram of a typical fiber gyro. The polarizer is positioned at the point where the light entering and leaving the interferometer follows a common path. The coupler immediately following the source does not have to have any special polarization properties as its function is to separate the light going into the interferometer from that returning to the detector. The parts of the interferometer following the polarizer need to preserve polarization to the maximum extent. The coil of elliptically cored fiber usually has an "h" factor of about 50 dBm so that a coil, say 1 km long, would hold polarization to 20 dB. The coupler made from D fiber (described in Chapter 7) preserves polarization to about 35 dB. The phase modulator is a disc of piezo-electric material with fiber wrapped around and fixed to the edge. A voltage applied across the disc at a frequency equal to the radial mechanical resonance of the disc applies pressure to the fiber to cause a change of index and hence of phase. The fiber components are spliced together using the fusion-splicing technique. The length between successive discontinuities (e.g., those due to components or splices) is made longer than the decoherence length mentioned in Chapter 7 in order to avoid coherent coupling of any out-of-polarization paths causing gyro drift.

The source at a wavelength of 815 nm is a laser diode below threshold in quasi-LED mode, which has sufficient line width and therefore short enough coherence length to make the decoherence length mentioned above a practical value. Typically the source line width is about 20 nm. The laser is pigtailed directly to the D fiber of the directional coupler.

Practical FOGs have to work over an extended temperature range, typically $-55°$ to

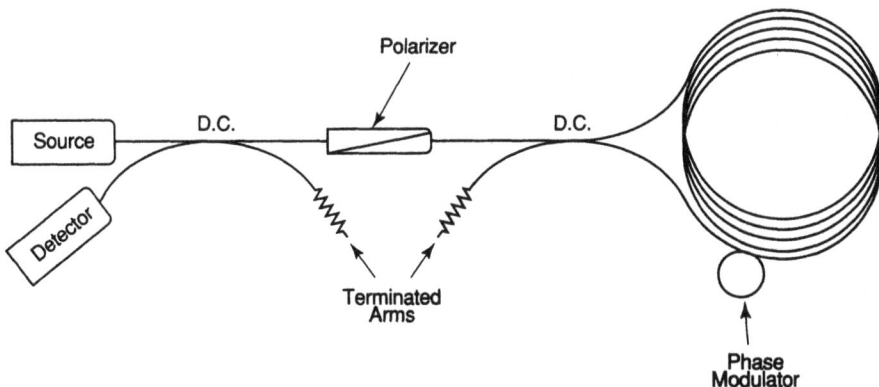

Figure 8.3 Schematic of fiber-optic gyro.

+85°C. It is essential that any material surrounding the fiber should not harden at low temperatures and cause degeneration of the polarization-preserving properties. The fiber itself has two coatings: a soft plastic inner coating, which maintains its pliability at low temperatures, surrounded by a hard outer coat for protection.

The elliptical core fiber has a distinct advantage when used in gyros in that, as can be seen from the figures given in Chapter 6, the variation of birefringence with temperature is only about 1/7 that of stress-induced birefringent fibers. Change of birefringence changes the state of polarization in the fiber, again causing problems of gyro drift.

A practical gyro built along the lines described above has the following characteristics and performance:

- Coil length of 755m, matched to modulation frequency;
- Phase modulation frequency of 136 kHz;
- Coil diameter of 0.08m;
- Source wavelength of 815 nm;
- Source linewidth of 20 nm;
- Detector current of 1 μA;
- Shot-noise limited rotation noise $0.14°/hr/\sqrt{H_z}$ (calculated from [4]);
- Measured rotation noise $0.22°/hr/\sqrt{H_z}$.

The elliptical core D fiber opens up the possibility of making a gyro with a spliceless Sagnac interferometer. A coil plus modulator is formed into a D-fiber coupler. The input fiber to the coupler is etched and coated with indium to form a polarizer. Such a system would eliminate spurious reflections and extra loss due to splices.

8.3 SENSORS USING THE HIGHER ORDER MODES IN ELLIPTICAL CORE FIBER

These sensors fall into the category of "properties of the elliptical core per se." The first set of four higher order modes in circular core fiber (the H_{01}, E_{01}, and the two HE_{21} modes with orthogonal transverse fields, which comprise the LP mode), have nearly identical cutoffs and propagation constants; that is, when the core-cladding index difference is small, as for instance in communication fibers (see Chapter 5). The circular geometry does not key the modes into any azimuthal position so that the transverse fields of the HE_{21} modes are free to move around with any slight perturbation of the fiber (see Figure 8.4). This causes instability in any devices that use the LP_{11} mode, such as intermodal couplers

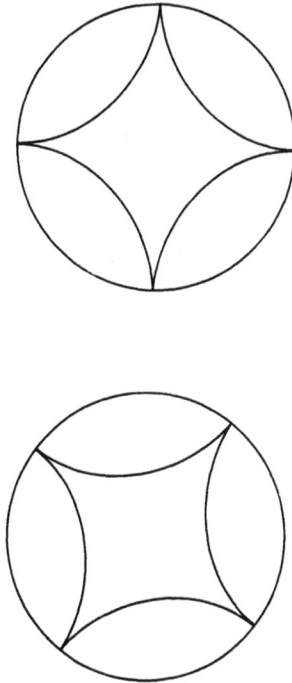

Figure 8.4 HE$_{21}$ mode in circular guide.

[5], modal filters [6], and acousto-optic single sideband frequency shifters [7]. Kim et al. [8] first proposed keying the modes and breaking the degeneracy with elliptical geometry. As has been shown in Chapter 5, the four modes now form two pairs, the HE$_{01}$ and EH$_{01}$, and the odd and even $_o$HE$_{21}$, $_e$HE$_{21}$ modes, each pair having the same number of azimuthal and radial variations of field (i.e., $_{01}$ and $_{21}$). For small index difference, each pair has virtually the same cutoff and propagation constant until the ellipticity becomes large. For large index difference, the modal pair splits at smaller ellipticities (Figure 5.6).

The proposal by Kim et al. [8] was to make a sensing interferometer using one of the fundamental modes $_o$HE$_{11}$ or $_e$HE$_{11}$ and the first LP$_{11}$ mode, comprising the HE$_{01}$ and EH$_{01}$ pair. With a coherent source the modes interfere, causing a change in the far-field radiation pattern of the fiber as the difference in modal propagation constants (the higher mode birefringence) changes with temperature, fiber stress under strain, or with acoustic effects. The setup is shown in Figure 8.5.

Blake et al. [9] in a companion paper to the seminal [8], describe the strain effects in two-mode elliptical core (2-mode E) fiber. The strain alters the beat length between the fundamental and higher order modes of the fiber caused by changes in core dimensions (via Poisson's ratio), index difference Δn and change in V. The varying beat length

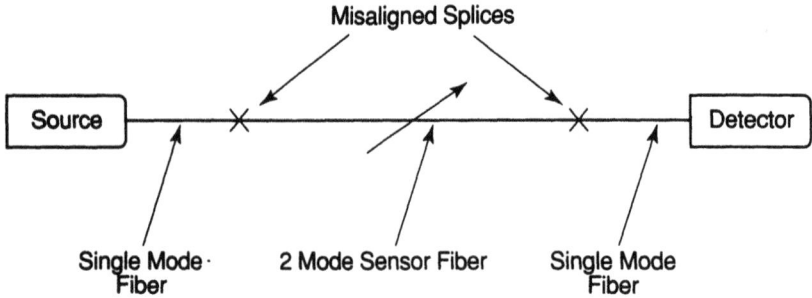

Figure 8.5 Overmoded E fiber sensor.

changes the far-field radiation pattern from the fiber. A typical beat length for a 2-mode E fiber, ellipticity $b/a = 0.5$, core cladding index difference $\Delta n = 0.035$ with fiber outer diameter 80 μm operated at $V_{op} = 1.85$, cutoff $V_c = 1.56$, $\dfrac{V_{op}}{V_c} \approx 1.18$ at a wavelength of 633 nm is 100 μm, compared with about 3 mm for the beat length between the fundamental modes at maximum birefringence below higher mode cutoff. The analysis in [9] was later extended by the same authors, Huang et al. [10], to include the separate effects of axial and radial strain, hydrostatic pressure, temperature, twisting, squeezing, and bending. A survey paper on the implementation methods for 2-mode E core fiber sensors has been written by Murphy et al. [11]. Application as vibration sensors on beams has been described by Vengsarkar et al. [12] and Murphy et al. [13], and for simultaneous measurements of strain and temperature (Michie et al. [14], and Vengsarker et al. [15].

Bohnert, de Wit and Nehring have described a remote voltage sensor using 2-mode E fiber strained by a piezoelectric element [16]. The sensor has two equal lengths of 2-mode E fiber, one acting as a remote sensing interferometer and the other as a local receiving interferometer, which has an additional piezoelectric element to adjust for any dc bias in phase caused, for instance, by change in temperature. The remote sensor is connected via two polarization maintaining fibers, with offset splices to the 2-mode E. The offset splices excite the fundamental and first higher order modes equally in order to get maximum interference. Figure 8.6 shows the arrangement. The minimum detectable phase shift quoted was less than 10^{-5} rad/$\sqrt{\text{Hz}}$.

8.4 BRAGG GRATINGS IN E FIBERS

Periodic variations, "gratings," in the characteristics of a fiber waveguide (e.g., index, diameter) can be used to couple modes. If the pitch of the variation is p, then coupling

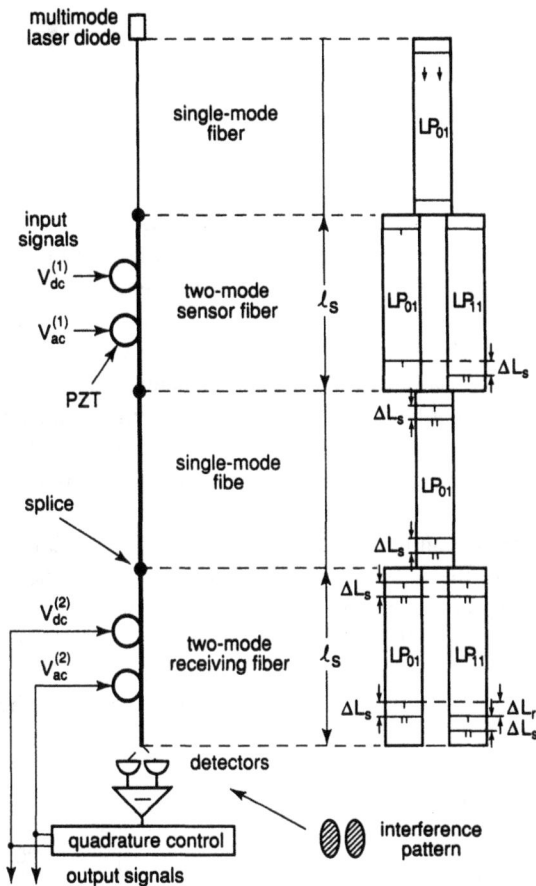

Figure 8.6 Remote voltage sensor using 2-mode elliptical fiber.

occurs if $\beta_p = \dfrac{2\pi}{p} = \beta_1 - \beta_2$. A familiar example is the coupling between the $_e\mathrm{HE}_{11}$ and $_o\mathrm{HE}_{11}$ modes by applying pressure to a fiber at a pitch equal to the beat length L_B

$$L_B = \frac{\lambda_0}{\Delta\bar{\beta}} = \frac{2\pi}{\Delta\beta} \quad ; \quad \frac{2\pi}{L_B} = \Delta\beta \tag{8.3}$$

If a traveling wave, propagation constant β, is reflected, the reflected wave has a propagation constant $-\beta$ so that

$$\beta_p = \beta - (-\beta) = 2\beta \tag{8.4}$$

then

$$\frac{2\pi}{p} = 2\beta = \frac{4\pi}{\lambda_g} \tag{8.5}$$

and the pitch p is equal to half the guide wavelength λ_g. Put another way, reflections caused by discontinuities that are spaced by half a guide wavelength add up in phase.

The most convenient way to produce periodic variations in the characteristics of a fiber is to change the index of the guiding region. This can be done by applied pressure. However, for reflection the pitch $p = \lambda_g/2$ is too small to be impressed laterally on the fiber, which is generally many wavelengths in diameter. An alternative way of changing index is by exposing the fiber to intense UV light (e.g., from an excimer laser at a wavelength of 249 nm), a phenomenon first discovered by Hill et al. in 1978 [17]. This photosensitivity is greater with fibers where the core contains a high concentration of germania.

A coupling grating can be formed inside a fiber by launching two modes that beat to form regions where the index is modified. This forms a kind of internal hologram where, after the grating is formed, launching one mode produces the other. Alternatively, the grating can be formed by illuminating the fiber sideways with a holographic interference pattern formed by two coherent beams (first suggested by Meltz, Morey, and Glen [18] in 1989, see Figure 8.7) or by using a mask to create a periodic illumination of the

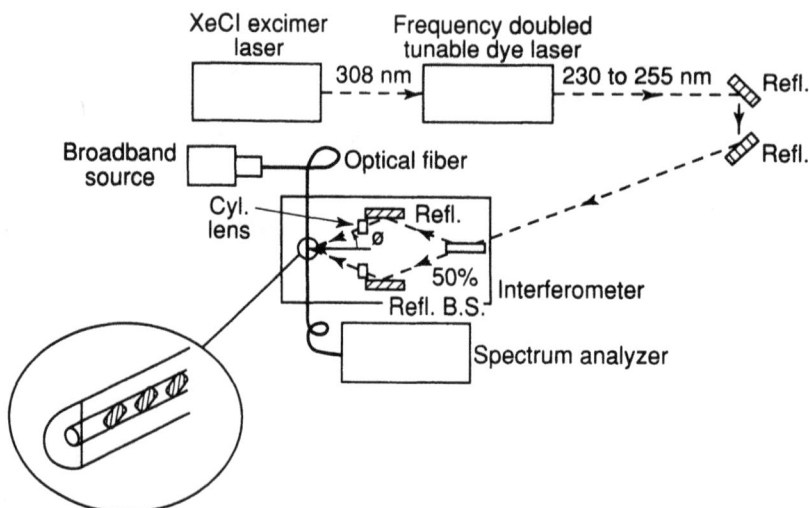

Figure 8.7 Holographic method of forming a grating in elliptical fiber [19].

fiber from one beam. Recently, gratings have been written into fibers during the drawing process. Very short powerful pulses from an excimer laser are used, either with the two-beam or the mask process to "freeze in" the periodicity after the fiber leaves the furnace and before it is coated (Askins et al. [19,20]).

Once formed, the fiber gratings have a number of uses. The grating pitch can be varied, for example, with temperature or strain, changing the wavelength of reflection and thus making a sensor. Attached to a semiconductor laser, a grating fiber with variable pitch (applied, say with a strain-inducing piezoelectric element) can tune the laser (see Figure 8.8, Mellis et al. [21], Morey, Meltz, and Glenn, [22]). Gratings can be used as mirrors on rare-earth-doped fiber lasers (see next chapter) where varying the grating pitch again changes the lasing wavelength. Vengsarkar, Greene, and Murphy [23] have made gratings in 2-mode E fiber using a high-intensity light ($\lambda = 514.5$ nm) from an argon ion laser. In such a fiber the beat length between the fundamental and first higher order modes becomes dependent on strain as the pitch of the coupling mechanism is changed. Such sensors have been used, among other things, to analyze the bending of beams.

8.5 COUPLING OF MODES USING ACOUSTIC WAVES

The fiber modes are coupled by using acoustic waves to generate pressure and therefore to change the propagation constant via a change in the index of the fiber material. The first experiments on elliptical core D fiber were reported by Risk et al. [24], who described the interaction between an acoustic wave traveling along the fiber and the fundamental $_e\text{HE}_{11}$ and $_o\text{HE}_{11}$ modes. The effect of the interaction is to couple one mode to another, the optical frequency being shifted by plus or minus the acoustic frequency in the process.

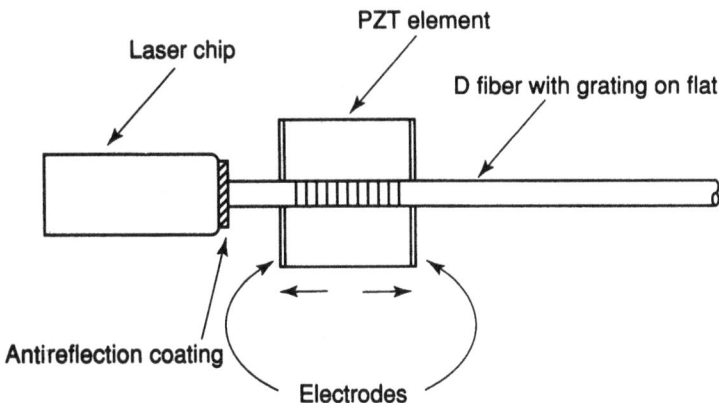

Figure 8.8 Laser tuning using grating on D fiber.

If the optical radian frequencies of the $_e\text{HE}_{11}$ and $_o\text{HE}_{11}$ modes are $_e\omega$ and $_o\omega$ (which are, of course, identical before coupling), the acoustic radian frequency is $_a\beta$ and the corresponding propagation constants are then $_e\beta$, $_o\beta$, and $_a\beta$ for strong coupling (since this is parametric coupling, with the material index as the parameter).

$$_e\omega = {_o\omega} \pm {_a\omega} \tag{8.6}$$

$$_e\beta = {_o\beta} \pm {_a\beta} \tag{8.7}$$

The phase velocity of the surface acoustic wave in silica

$$_av = \frac{_a\omega}{_a\beta} \tag{8.8}$$

will not fit these conditions exactly so that the acoustic wave (traveling in the material in contact with the fiber) has to propagate at an angle ϕ to the fiber axis. Then

$$_e\beta = {_o\beta}a \pm \cos\phi$$

with $\Delta\beta = \frac{2\pi}{\lambda_0}\overline{\Delta\beta}$ and the beat length $L_B = \frac{\lambda_0}{\Delta\beta}$

$$_a\beta\cos\phi = \frac{2\pi}{_a\lambda} = \frac{2\pi}{L_B} ; \; _a\lambda = L_B \tag{8.9}$$

or the effective acoustic wavelength $_a\lambda$ in the fiber is equal to the beat length L_B.

If the acoustic wave is traveling in the same direction as the light in the fiber, the fast (even) wave will be coupled to the slow (odd) wave with an increase in frequency and the slow wave will be coupled to the fast wave with a decrease in frequency. If the acoustic wave is traveling against the optical wave, the fast wave is increased in frequency and the slow wave decreased.

For optimum coupling, the axes of the core ellipse must be at 45° to the applied stress so that both modes are coupled equally. Figure 8.9 shows the experimental arrangement of Risk et al. [24]. Because the acoustic wavelength is equal to the beat length, which in typical elliptical core fiber over the range of the usual optical wavelengths is a few millimeters, surface wave mode coupling/frequency shifting is limited to a few MHz.

An alternative scheme, also described in [24] overcomes this difficulty by using bulk acoustic waves launched laterally and meeting the fiber axis at an angle Θ, which can be adjusted so that the velocity of the wave moving along the fiber matches the beat length. Figure 8.10 from [24] shows the arrangement. The effect is analogous to the velocity along the shoreline of an angled wave breaking on a beach. The velocity of the wave along the fiber can be made infinite by decreasing Θ to 0. A more recent paper by Sevic and Patterson [25] describes modal coupling in D fiber using bulk-shear acoustic waves with coupling efficiencies of 14.3%, a center frequency of 75 MHz with a 3-dB bandwidth of 3.6 MHz.

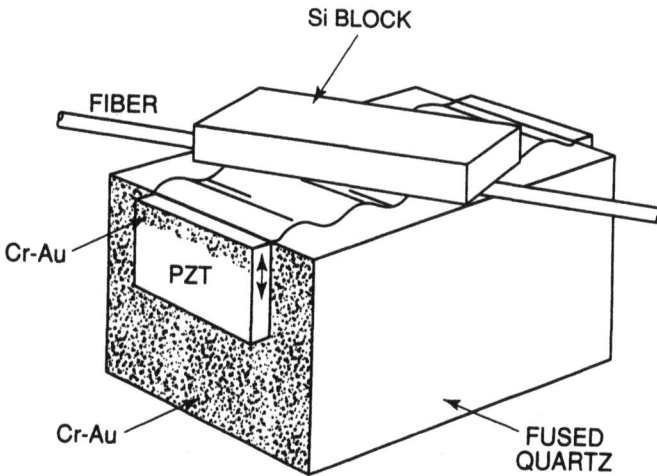

Figure 8.9 Surface acoustic wave coupling on D fiber.

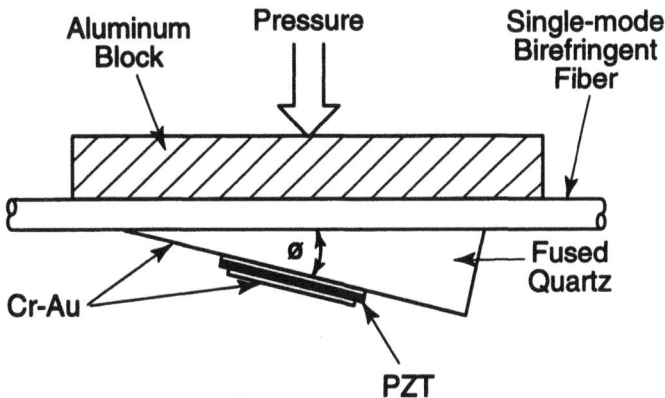

Figure 8.10 Bulk acoustic wave coupling on D fiber.

8.6 USE OF THE OPTICAL KERR EFFECT

The optical Kerr effect produces a change in the birefringence of a fiber via a change in refractive index of the guiding material in the presence of a strong optical electric field, produced for instance, by a high-power "pump" wave. The change in birefringence is proportional to the square of the amplitude of the peak electric field E volts/meter, which is related to the optical power density I of the pump.

$$I = \frac{E^2}{Z_w} \text{ watts/m}^2 \tag{8.10}$$

where Z_w is the waveguide impedance of the fiber

$$Z_w = \frac{120\pi}{\bar{\beta}} \text{ ohms} \tag{8.11}$$

so that

$$E^2 = IZ_w = I\frac{120\pi}{\bar{\beta}} \left(\frac{\text{volts}}{\text{m}}\right)^2 \tag{8.12}$$

Assuming that most of the power is concentrated in the fiber core with cross-sectional area $\pi\, ab$ (where a and b are the semimajor and semiminor axes of the core ellipse), then for a pump power of W_p watts

$$I = \frac{W_p}{\pi ab} \tag{8.13}$$

$$E^2 = \frac{W_p}{\pi ab} \cdot \frac{120\pi}{\bar{\beta}} = \frac{120 W_p}{ab\bar{\beta}} \left(\frac{\text{volts}}{\text{m}}\right)^2 \tag{8.14}$$

for example, for a pump power $W_p = 1$ watt and typical core dimensions $2a \times 2b = 2 \text{ μm} \times 1 \text{ μm}$ with $\bar{\beta} = 1.46$

$$E^2 = \frac{120}{1.46 \times 0.5 \times 10^{-12}} = 1.64 \times 10^{14} \left(\frac{\text{volts}}{\text{m}}\right)^2$$

$$E = 1.28 \times 10^7 \left(\frac{\text{volts}}{\text{m}}\right)$$

Cokgor, Handerek, and Rogers [26] have described a method of using the optical Kerr effect to find points along an elliptical core fiber where there is coupling between modes, caused, for instance, by pressure applied locally. A high-power pulse at the pump wavelength is launched along one of the fiber axes. As it travels down the fiber, it produces additional birefringence, giving the effect of a short moving beat pattern. A probe signal is launched traveling in the opposite direction to the pump. Coupling between the modes of the probe signal is caused by the traveling beat pattern, which also changes any local coupling caused by pressure points. The position of these points can then be found from the timing of the moving beat pattern.

An optical Kerr switch using two-mode elliptical core fiber has been described by Park, Pohalski, and Kim [27]. Here the phase difference between a fundamental mode and the first pair of higher order modes is moved by the change in guide index produced by a high-power pump working in the fundamental mode region. The initial modal phase difference is tuned by stretching the fiber until there is one bright and one dark lobe in the far-field radiation pattern. Turning on the pump switches the lobes. Figure 8.11 shows the

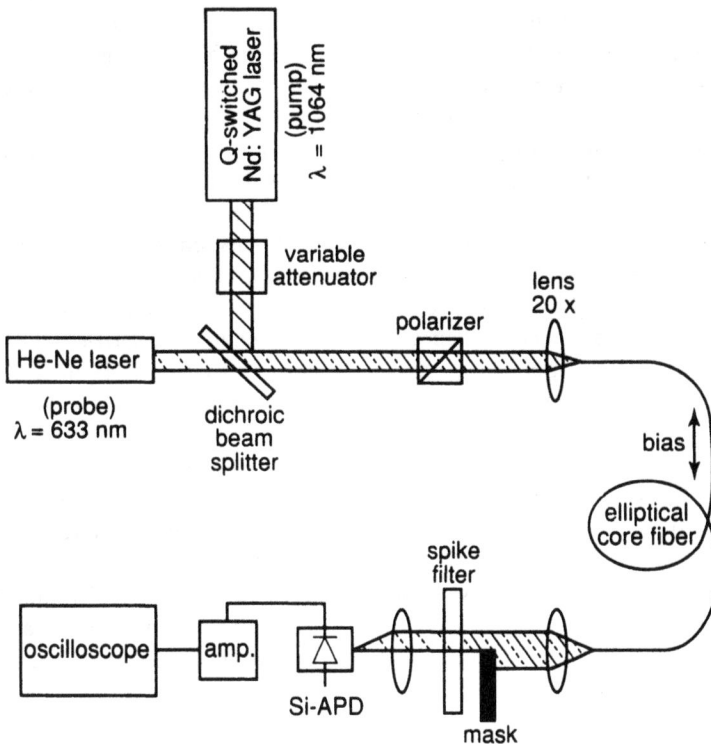

Figure 8.11 Switching using the optical Kerr effect.

experimental arrangement. The peak power at 1,064 nm, needed to switch at 633 nm, was reported to be 1.6W. A variation on this theme using periodic coupling between counterpropagating pump and signal waves has been reported by Park, Huang, and Kim [28].

A. S. Davison and I. H. White [29] have described an optical amplifier using the Kerr effect. A pump is launched into the fiber polarized at 45° to the birefringent axes so as to excite each fundamental mode equally. A signal is coupled into one of the fiber modes and causes a phase shift between modes due to the Kerr effect. The induced phase shift difference rotates the polarization of the pump. The rotation in polarization is then converted to amplitude modulation via a polarizer. Thus the modulation at the signal wavelength causes modulation at the high-power pump wavelength with a combined amplification and change in frequency.

8.7 THE ELLIPTICAL CORE FIBER AS A CURRENT SENSOR

Many schemes have been proposed whereby the Faraday effect, rotating the polarization in a single-mode fiber with an axial magnetic field, is used to measure current in a conductor. Such a device would be ideal for monitoring high-voltage transmission lines where the conventional current transformers are bulky and expensive to construct because of the problem of voltage breakdown across the magnetic coupling.

The proposed schemes generally use a coil of single-mode circular cored fiber wrapped around the high-voltage conductor so that the circumferential magnetic field surrounding the conductor acts along the axis of the fiber to rotate the polarization of the fundamental HE_{11} mode. Polarized light is delivered to this sensing fiber via a polarization-holding fiber from a light source at ground potential, the length of fiber providing an ideal insulating path. The light with rotated polarization is returned to a polarization analyzer and detector at ground potential by another insulating length of polarization-holding fiber. A survey of the many variations on this theme is given by Rogers [30].

The universal problem is that in order to rotate the polarization of the HE_{11} mode, which is equivalent to coupling power from a mode with the transverse field in one direction to one where the field is orthogonal to that direction, both modes must have identical propagation constants; otherwise, the fields of the two modes would go in and out of phase transferring power to and fro. A fiber with a perfectly circular core would have equal propagation constants for orthogonal HE_{11} modes, but there is always some noncircularity resulting in slight differences in propagation constant. Attempts have been made to overcome this problem by spinning the fiber as it is drawn, thus averaging out any noncircularity. However, winding the fiber into a coil introduces stress causing birefringence which varies with temperature and vibration.

A solution to the problem, proposed by Stolen [31], is to use highly birefringent fiber and to apply a periodic axial magnetic field in such a way that the field is only present when the waves in the two orthogonal modes (now traveling at very different phase velocities) are in phase. The period, then, is equal to the beat length. Such a periodic field could be produced, for instance, by spaced magnetic screening along the fiber. Figure 8.12 shows a schematic of the arrangement. Experiments with periodic magnetic coupling using elliptically cored fiber have been reported by Chu, McStay, and Rogers [32].

A remaining problem is the need to match the magnetic periodicity with the beat length of the fiber over lengths long enough to give a practical amount of coupling and over the operative range of temperatures. Some advantage can be had by ''broadbanding'' the structure with a pitch that varies along the coupling length. Another technique is to broadband the source, thus creating a spread in beat length L_B that is related directly to wavelength.

$$L_B = \frac{\lambda_0}{\overline{\Delta\beta}}$$

8.8 CORRECTION OF DISPERSION USING OVERMODED FIBER

The examples given so far of the use of overmoded elliptical core fiber have been related to sensors in some form or other. In contrast, an interesting application for correcting the effects of dispersion in long-distance transmission lines has been proposed by Poole, Wiesenfeld, McCormick, and Nelson [33]. The correction depends on the fact that towards cutoff, the first higher order modes (the H_{01} and E_{01} in circular fiber or the $_eHE_{01}$ and $_oEH_{01}$ pair in elliptical fiber) have a group velocity that increases with wavelength because as cutoff is approached, more of the power is carried in the lower index cladding due to the expansion of the evanescent field. This increasing group velocity with wavelength produces a dispersion that is opposite in sign to that of conventional communication fiber

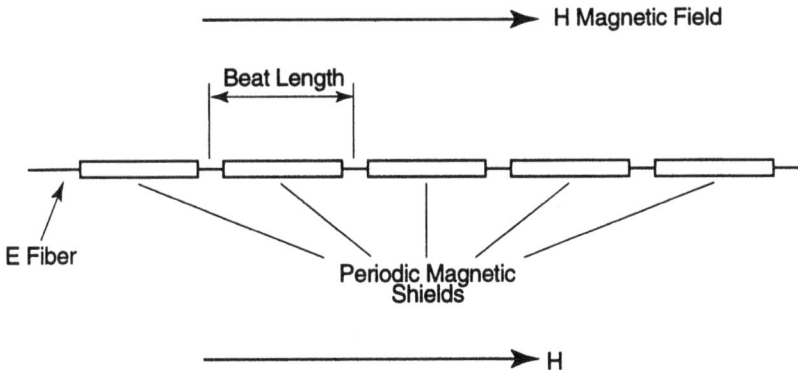

Figure 8.12 Periodic Faraday rotation as a current sensor.

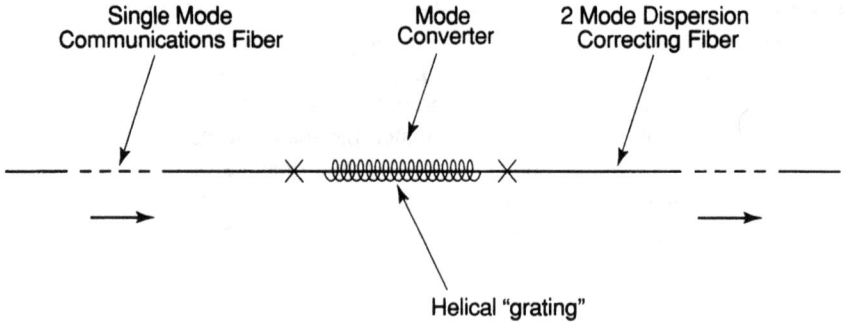

Figure 8.13 Correction of dispersion using overmoded elliptical fiber.

at 1,500 nm. Furthermore, because the rate of change of the power distribution from core to cladding itself increases towards cutoff, the second order group velocity dispersion is also opposite in sign to that of the single-mode fiber.

In the experimental realization of this idea, described in [33], the power in the fundamental mode in the communication fiber is converted to the first higher order modes in an overmoded fiber by means of a helical grating that introduces microbending of the fiber at such a pitch as to couple the propagation constants. The helical grating coupler (also described in a prior experiment [34]) consists of 20 turns of 243-μm gold wire wound directly onto the fiber, which gives an energy conversion between the fundamental and higher order modes of 96%. A schematic of the idea is shown in Figure 8.13. The dispersion of the combined communication fiber and dispersion-correcting fiber is reduced to less than 0.4 ps/(nm·km) over the entire working band, 1,530 to 1,560 nm.

REFERENCES

[1] Culshaw, B. and J. Dakin, Editors. *Optical Fiber Sensors,* Vols. 1 and 2, Artech House: Boston: 1988 and 1989.

[2] Lefèvre, H. *The Fiber-Optic Gyroscope,* Artech House: Boston: 1993 pp. 285–299.

[3] Kintner, E. C. "Polarization Control in Optical-Fiber Gyroscopes," *Optics Letters,* Vol. 6, 1981, pp. 154–156.

[4] Burns, W. K., R. P. Moeller, and A. Dandridge. "Excess Noise in Fiber Gyroscope Sources," *IEEE Photonics Tech. Letter,* Vol. 2, 1990, p. 606.

[5] Blake, J. N., B. Y. Kim, and H. J. Shaw. "Fiber Optic Modal Coupler Using Periodic Microbending," *Optics Letters,* Vol. 11, 1986, p. 177.

[6] Sorin, W. V., B. Y. Kim, and H. J. Shaw. "Highly Selective Evanescent Modal Filter for 2 Mode Optical Fibers," *Optics Letters,* Vol. 11, 1986, p. 581.

[7] Kim, B. Y., J. N. Blake, H. E. Engan, and H. J. Shaw. "All-fiber acousto-optic frequency shifter," *Optics Letters,* Vol. 11, 1986, p. 389.

 [8] Kim, B. Y., J. H. Blake, S. Y. Huang, and H. J. Shaw. "Use of Highly Elliptical Core Fibers for Two-Mode Fiber Devices." *Optics Letters*, Vol. 12, 1987, p. 729.

 [9] Blake, J. H., S. Y. Huang, B. Y. Kim, and H. J. Shaw. "Strain Effects in Highly Elliptical Core Two-Mode Fibers," *Optics Letters*, Vol. 12, 1987, p. 732.

[10] Huang, S. Y., J. H. Blake, and B. Y. Kim. "Perturbation Effects on Mode Propagation in Highly Elliptical Core Two-Mode Fibers," *Journal Lightwave Technology*, Vol. 8, 1990, p. 23.

[11] Murphy, K. A., M. S. Miller, A. M. Vengsarkar, and R. O. Claus. "Elliptical-Core Two-Mode Optical-Fiber Sensor Implementation Methods." *Journal Lightwave Technology*, Vol. 8, 1990, p. 1688.

[12] Vengsarkar, A. M., B. R. Fogg, W. V. Miller, K. A. Murphy, and R. O. Claus. "Elliptical-Core Two-Mode Optical Fiber Sensors as Vibration-Mode Sensors," *Electronics Letters*, Vol. 27, 1991, p. 931.

[13] Murphy, K. A., B. R. Fogg, and A. M. Vengsarkar. "Spatially Weighted Vibration Sensors Using Tapered Two Mode Optical Fibers." *Journal Lightwave Technology*, Vol. 10, 1992, pp. 1680–1687.

[14] Michie, W. C., B. Culshaw, S. J. Roberts, and R. Davidson. "Fiber Optic Technique for Simultaneous Measurement of Strain and Temperature," *Proc. SPIE*, Vol. 1588, 1991, p. 342.

[15] Vengsarkar, A. M., C. Michie, J. Ljilja, B. Culshaw, & R. O. Claus. "Fiber-Optic Dual-Technique Sensor for Simultaneous Measurement of Strain and Temperature," *Journal Lightwave Technology*, Vol. 12, 1994, p. 170.

[16] Bohnert, K., G. C. de Wit, and J. Nehring. "Interferometric Dual Mode Fiber Voltage Sensor with Remote Coherence—Tuned Interrogation," *Proc. 9th Optical Fiber Sensor Conference*, Firenze, Italy, 1993.

[17] Hill, K. O., Y. Fujii, D. C. Johnson, and B. S. Kawasaki. "Photosensitivity in optical fiber waveguides: Application to reflection filter fabrication," *Applied Physics Letters*, Vol. 32, 1978, p. 647.

[18] Meltz, G., W. W. Morey, and W. H. Glenn. "Formation of Bragg Gratings in Optical Fibers by a Transverse Holographic Method," *Optics Letters*, Vol. 14, 1989, p. 823.

[19] Askins, C. G., T. E. Tsai, G. M. Williams, M. A. Putnam, M. Bashkansky, and E. J. Friebele. "Fiber Bragg Reflectors Prepared by a Single Excimer Pulse," *Optics Letters*, Vol. 17, 1992, p. 833.

[20] Askins, C. G., M. A. Putnam, G. M. Williams, and E. J. Friebele. "Optical Fiber Bragg Gratings Produced During Fiber Draw," *Proc. SPIE*, Vol. 2044, 1993.

[21] Mellis, J., S. A. Al-Chalabi, K. H. Cameron, R. Wyatt, J. C. Regnault, W. J. Devlin, and M. C. Brain. "Miniature Packaged External-Cavity Semi Conductor Laser with 50 GHz Continuous Electrical Tuning Range," *Electronics Letters*, Vol. 24, August 1988, p. 988.

[22] Morey, W. W., G. Meltz, and W. H. Glenn. "Fiber Optic Bragg Grating Sensors," *Proc. SPIE*, Vol. 1169, 1989, p. 98.

[23] Vengsarkar, A. M., J. H. Greene, and K. A. Murphy. "Photo Induced Refractive-Index Changes in Two-Mode Elliptical Core-Fibers: Sensing Applications," *Optics Letters*, Vol. 16, 1991, p. 1541.

[24] Risk, W. P., R. C. Youngquist, G. S. Kino, and H. J. Shaw. "Acousto-Optic Frequency Shifting in Bi-Refringent Fiber," *Optics Letters*, Vol. 9, 1984, p. 309.

[25] Sevic, J. F., and D. B. Patterson. "Non-Invasive Polarization Eigenmode Coupling in Elliptical Core Optical Fiber Based on Bulk-Shear Acoustic Waves," *Optics Letters*, Vol. 18, 1993, p. 2008.

[26] Cokgor, I., V. A. Handerek, and A. J. Rogers. "Distributed Optical-Fiber Sensor for Spatial Location of Mode Coupling by Using the Optical Kerr Effect," *Optics Letters*, Vol. 18, 1993, p. 705.

[27] Park, H. G., C. Pohalski, and B. Y. Kim. "Optical Kerr Switch Using Elliptical-Core Two-Mode Fiber," *Optics Letters*, Vol. 13, 1988, p. 776.

[28] Park, H. G., S. Y. Huang, and B. Y. Kim. "All-Optical Intermodal Switch Using Periodic Coupling in a Two Mode Waveguide," *Optics Letters*, Vol. 14, 1989, p. 877.

[29] Davison, A. S., and I. H. White. "Initial Demonstration of a Novel Broadband Optical Amplifier Using the Kerr Effect in an Optical Fiber," *Optics Letters*, Vol. 14, 1989, p. 802.

[30] Rogers, A. J. "Optical-Fiber Current Measurement," *International Journal of Opto Electronics*, Vol. 3, 1988, p. 391.

[31] Stolen, R. H., and E. H. Turner. "Faraday Rotation in Highly Bi-Refringent Optical Fibers," *Applied Optics*, Vol. 19, 1980, p. 842.

[32] Chu, W., D. McStay, and A. J. Rogers. "Current Sensing by Mode Coupling in Fiber Via the Faraday Effect," *Electronics Letters,* Vol. 27, 1991, p. 207.

[33] Poole, C. D., J. M. Wiesenfeld, A. R. McCormick, and K. T. Nelson. "Broadband Dispersion Compensation by Using the Higher Order Spatial Mode in a Two-Mode Fiber," *Optics Letters,* Vol. 17, 1992, p. 985.

[34] Poole, C. D., C. D. Townsend, and K. T. Nelson. "Helical-Grating Two-Mode Fiber Spatial-Mode Coupler," *Journal Lightwave Technology,* Vol. 9, 1991, p. 598.

Chapter 9

Rare-Earth-Doped Elliptically Cored Fiber

9.1 BACKGROUND

The first rare-earth-doped glass fiber was demonstrated by Koester and Snitzer in 1963 [1], well before the birth of optical-fiber communications in 1966 [2]. The core, 10 µm in diameter, was doped with neodymium and surrounded by a large diameter (0.75 to 1.5 mm) cladding. The whole thick fiber, 1m long, was wound into a helix by heating until it softened. It was pumped by a flash tube along the axis of the helix, and gave gains of up to 47 dB at a wavelength of 1.06 µm.

The first end-pumped fiber laser was reported by Stone and Burris [3] ten years later in 1973, but it was not until the publication in 1985 by Mears et al. [4] of a description of the first single-mode Nd-doped low-loss fiber laser that interest quickened with the realization that fiber amplifiers could be used to bolster the signal in optical-fiber communication systems. Since then the subject has exploded and indeed has been described as the most important development since optical-fiber communication itself.

There are many good books on the subject. A particularly useful publication is one of the *SPIE Milestone Series,* which binds together about 170 selected papers on fiber sources and amplifiers [5].

This chapter concentrates on the rare-earth-doped elliptical core fiber, which has some unique properties.

9.2 FABRICATION

There are several ways of introducing the rare-earth dopant into the fiber. An effective method for the elliptically cored fiber manufactured by the process described in Chapter 6 is known as solution doping [6] and is illustrated in Figure 9.1.

The cladding material (fluorine-doped silica) is deposited in the usual way using the MCVD process. The germania core is then deposited, but at a reduced temperature so that

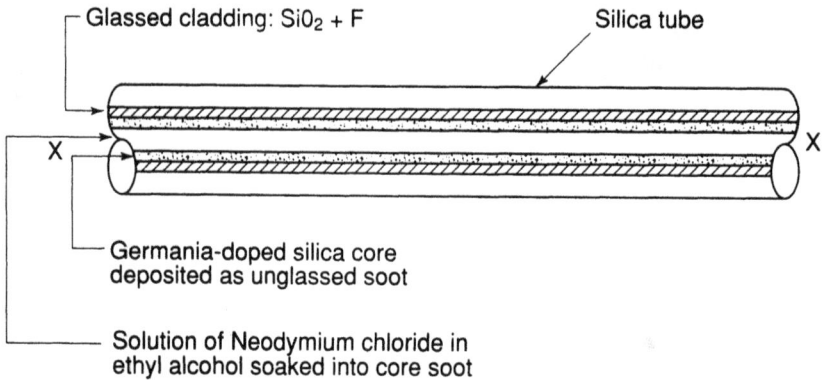

Figure 9.1 Solution-doping method for making preforms.

it does not form a glass but remains as a porous soot on the walls of the tube. A solution of a salt of the rare earth (usually a chloride or nitrate) in alcohol or water is introduced into the tube and soaks into the porous soot, which is then glassed at a higher temperature, after which the tube is collapsed to form the preform. Passing chlorine through the heated tube prior to glassification reduces the OH content considerably. The concentration of the rare-earth dopant is adjusted by diluting the saturated salt solution by the required amount.

9.3 FIBER CHARACTERISTICS

Figure 9.2 shows the superluminescent spectrum of a fiber made by the solution doping method with neodymium chloride dissolved in 100% ethyl alcohol at a dilution of 1 in 500 from saturation. The length of the fiber is 25m. It is pumped at a wavelength of 812 nm to give 37 μW of power. Figure 9.3 shows the spectrum of 50m of fiber doped at 1 in 1,000 and pumped at the same wavelength to give 49 μW. Figure 9.4 shows the spectrum, expanded about the main transition at 1,080 nm of 9m of fiber doped at 1 in 500. It shows a linewidth, between the markers at half-peak power, of 68 nm. Figure 9.4(a) shows the attenuation of the pump power as a function of the concentration of the dopant solution. Nd doped fibers such as these have been used as sources for fiber-optic gyroscopes. They have the advantages of an output wavelength that is insensitive to temperature when compared with other sources such as superluminescent diodes and of being relatively inexpensive since the pump can be a cheap laser such as those used in compact disc players.

9.4 LOOP MIRROR REFLECTORS

The efficiency of Nd-doped fiber superluminescent sources is improved by pumping through a loop mirror (see Chapter 7), which can be made to act as a 100% reflector at

RL 258.2 nW
SENS 270 pW
LINEAR

START 800.0 nm
* RB 1 nm   VB 2 kHz   S

STOP 1500.0 nm
ST 1.8 sec

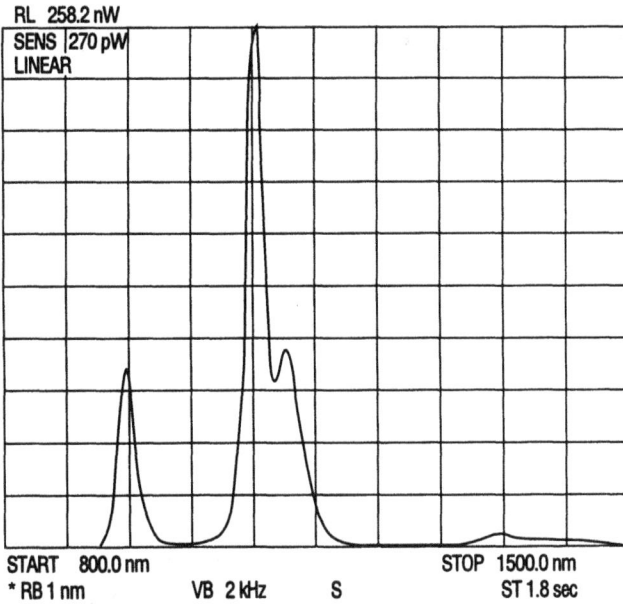

Figure 9.2 Superluminescent spectrum of fiber-doped 1 in 500 dilution from saturated solution.

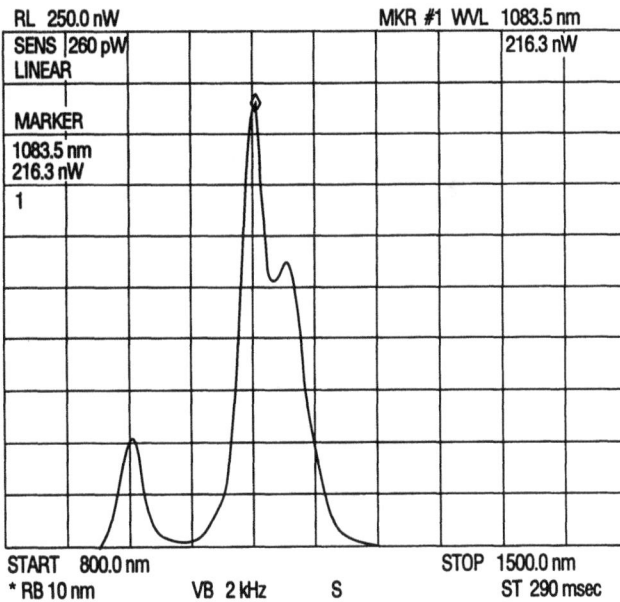

RL 250.0 nW   MKR #1 WVL 1083.5 nm
SENS 260 pW   216.3 nW
LINEAR

MARKER
1083.5 nm
216.3 nW

1

START 800.0 nm   STOP 1500.0 nm
* RB 10 nm   VB 2 kHz   S   ST 290 msec

Figure 9.3 Superluminescent spectrum of fiber-doped 1 in 1,000 dilution from saturated solution.

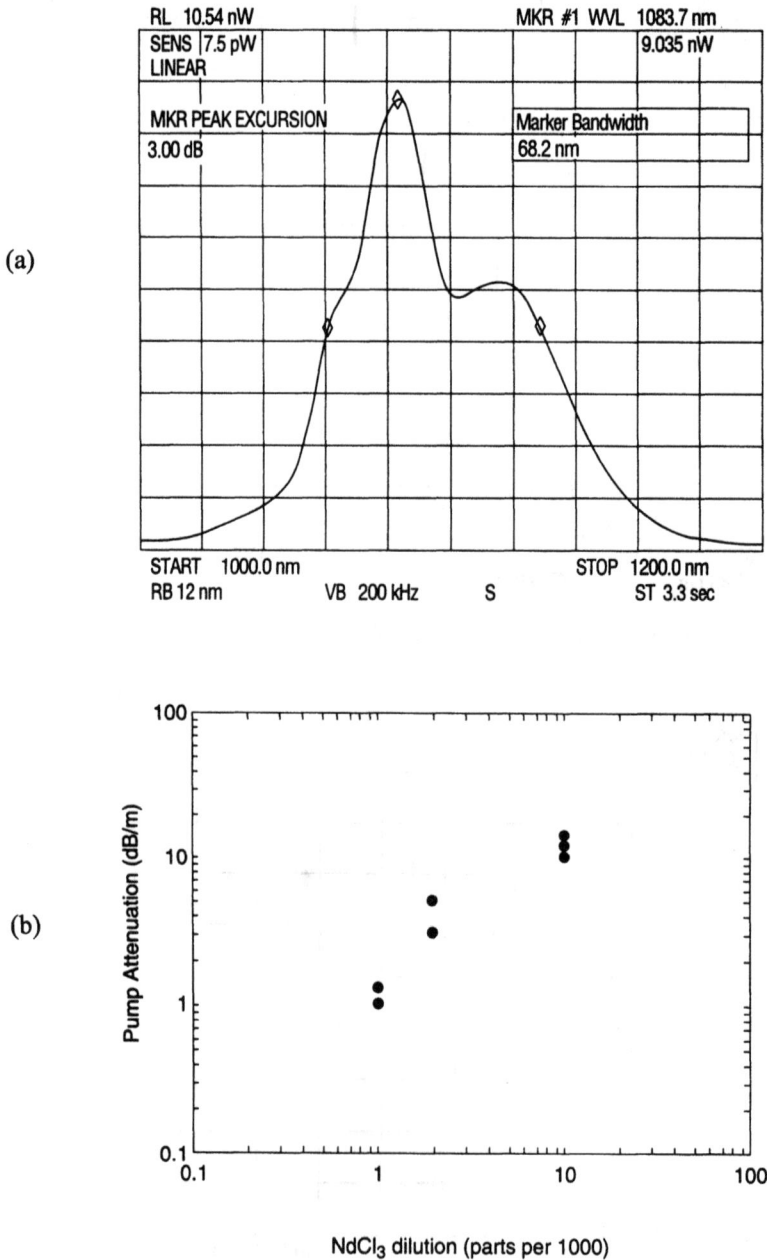

Figure 9.4 (a)Expanded spectrum of fiber doped at 1 in 500 dilution from saturated solution. (b)Attenuation of the pump power as a function of concentration of dopant solution.

1080 mm while passing the pump at 812 nm with no reflection (see Figure 9.5). This is particularly easy to do with the evanescent field D-fiber coupler described in Chapter 7 because the field which couples at 1080 nm is tightly confined and does not couple at 812 nm. With such an arrangement, light in the 1080-nm band is spontaneously emitted and travels in both directions along the fiber. Light traveling in the direction reverse to that of the pump is reflected from the loop and is then amplified to augment that traveling with the pump. Figure 9.6 is the spectrum of a Nd-doped D fiber plus loop mirror, showing the enhanced output at 1080 nm relative to that at 940 nm compared with the spectrum of the arrangement without a loop mirror shown in Figure 9.3.

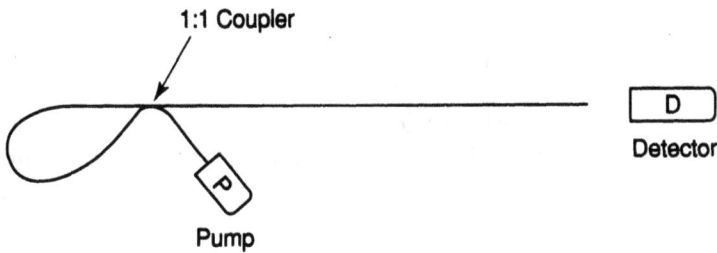

Figure 9.5 Pumping through a loop mirror.

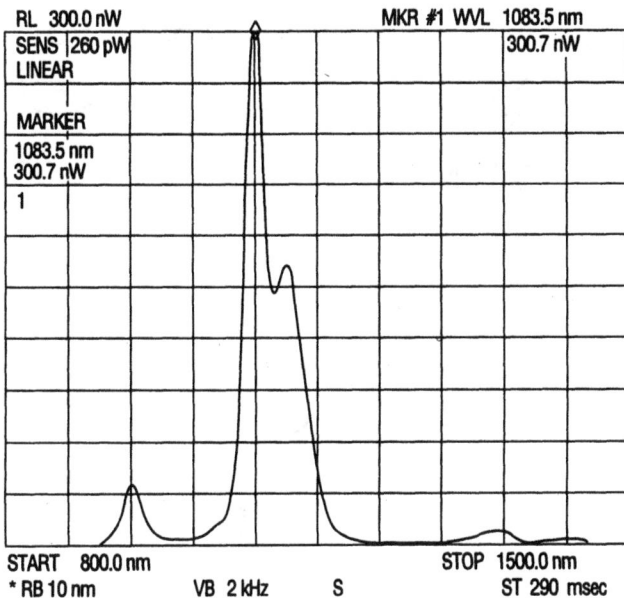

Figure 9.6 Spectrum of Nd doped fiber plus loop mirror.

Fiber lasers using loop mirrors were first reported by Miller et al. in 1987 [7]. Here there is a loop on each end of the fiber, the pump being fed through one of the loops and the laser output taken from the other. The reflection coefficient of the output loop is adjusted via the degree of coupling in the loop, a 1:1 ratio giving complete reflection (see Figure 9.7).

9.5 FIBER LASERS WITH ELLIPTICAL CORES

Srinivasan et al. [8] have studied the polarization properties of Nd-doped elliptical core fiber and found differential gain between the $_eHE_{11}$ and $_oHE_{11}$ modes, the gain of the latter being greater than that of the former. The experimental arrangement is shown in Figure 9.8. Approximately 1m lengths of circular and elliptical core fiber doped with Nd at approximately 100 ppm were compared for polarization aspect ratio; that is, for the ratio of powers in orthogonal polarizations (along the major and minor axes for the elliptical core). The elliptical core fiber had an index difference $\Delta n = 0.032$ and a birefringence $\Delta \bar{\beta} = 4 \times 10^{-4}$ with ellipticity $b/a = 0.5$. The circular core fiber had identical character-

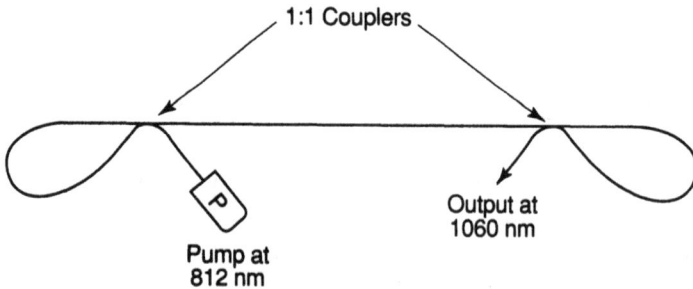

Figure 9.7 Fiber laser using loop mirror reflectors.

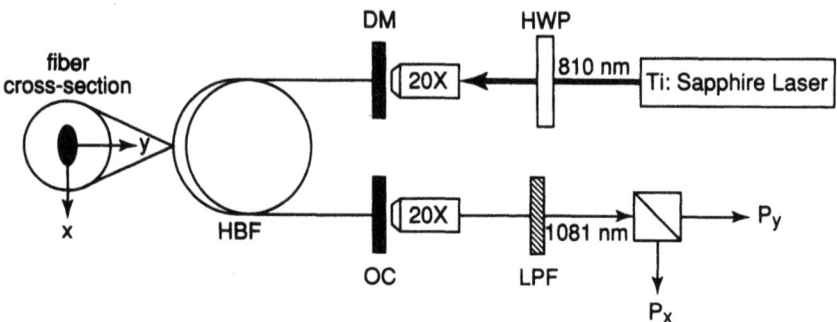

Figure 9.8 Differential gain experiment.

istics except, of course, with $b/a = 1.0$. The elliptical core fiber held the polarization of the pump (810 nm) when launched on one of the birefringent axes. When pumped with circularly polarized light, the polarization aspect ratio of the elliptical core fiber

$$H = \frac{P_{o\text{HE}_{11}}}{P_{e\text{HE}_{11}}} \approx 2.5$$

whereas for the circular core fiber

$$H = 1$$

as expected.

Figure 9.9 shows H as a function of input angle for a linearly polarized pump for two reflection coefficients R at the output of the laser $R = 0.5$ and $R = 0.9$.

An explanation of the difference in gain between the two fundamental modes may

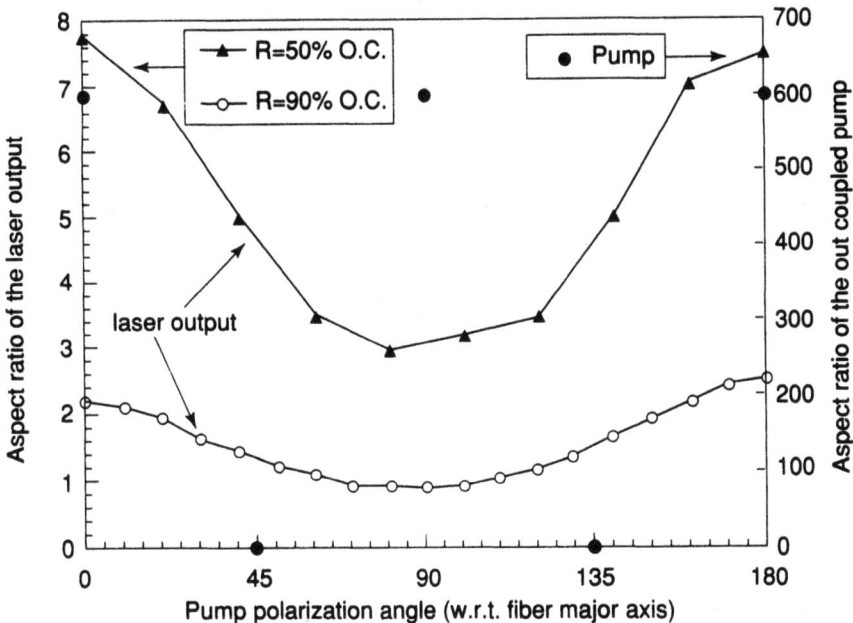

Figure 9.9 Aspect ratio as a function of pump polarization angle.

be that the fraction of power carried in the core, $\eta = \dfrac{P_{core}}{P_{total}}$, is slightly greater for the $_oHE_{11}$ mode than for the $_eHE_{11}$ mode (see Chapter 4, Figures 4.12, 4.13). Equation (2.15) gives

$$\eta = \frac{\dfrac{\bar{\beta}}{\bar{v}_g} - n_2^2}{n_1^2 - n_2^2} \tag{9.1}$$

where $\bar{\beta} = \dfrac{\lambda_0}{2\pi}\beta$ and $\bar{v}_g = \dfrac{v_g}{c}$; $\bar{\beta}$ is the normalized constant and $\bar{v}_g$ the group velocity v_g normalized to c. The difference between the confinement of power in the two modes is

$$\Delta\eta = {}_o\eta - {}_e\eta = \frac{\dfrac{_e\bar{\beta}}{_e\bar{v}_g} - n_2^2 - \left(\dfrac{_o\bar{\beta}}{_o\bar{v}_g} - n_2^2\right)}{n_1^2 - n_2^2} \tag{9.2}$$

which reduces to

$$\Delta\eta = {}_o\eta - {}_e\eta = \frac{_o\bar{v}_g\,_e\bar{\beta} - {}_e\bar{v}_g\,_o\bar{\beta}}{(n_1^2 - n_2^2)\,\bar{v}_g^2} \tag{9.3}$$

so that the fractional change in confinement

$$\frac{\Delta\eta}{\eta} = \frac{_o\bar{v}_g\,_e\bar{\beta} - {}_e\bar{v}_g\,_o\bar{\beta}}{\bar{v}_g\bar{\beta} - n_2^2\bar{v}_g^2} \tag{9.4}$$

Using computed values then to a good approximation (about 1% for the usual range of index difference Δn in fibers)

$$\frac{\Delta\eta}{\eta} \propto \Delta n$$

Figure 9.10 shows $\dfrac{\Delta\eta}{\eta\Delta n}$, the fractional change in confinement normalized with respect to Δn for three ellipticities, $b/a = 0.5, 0.7$, and 0.8. The normalization of $\dfrac{\Delta\eta}{\eta}$ with respect to Δn is easily verifiable for the slab waveguide.

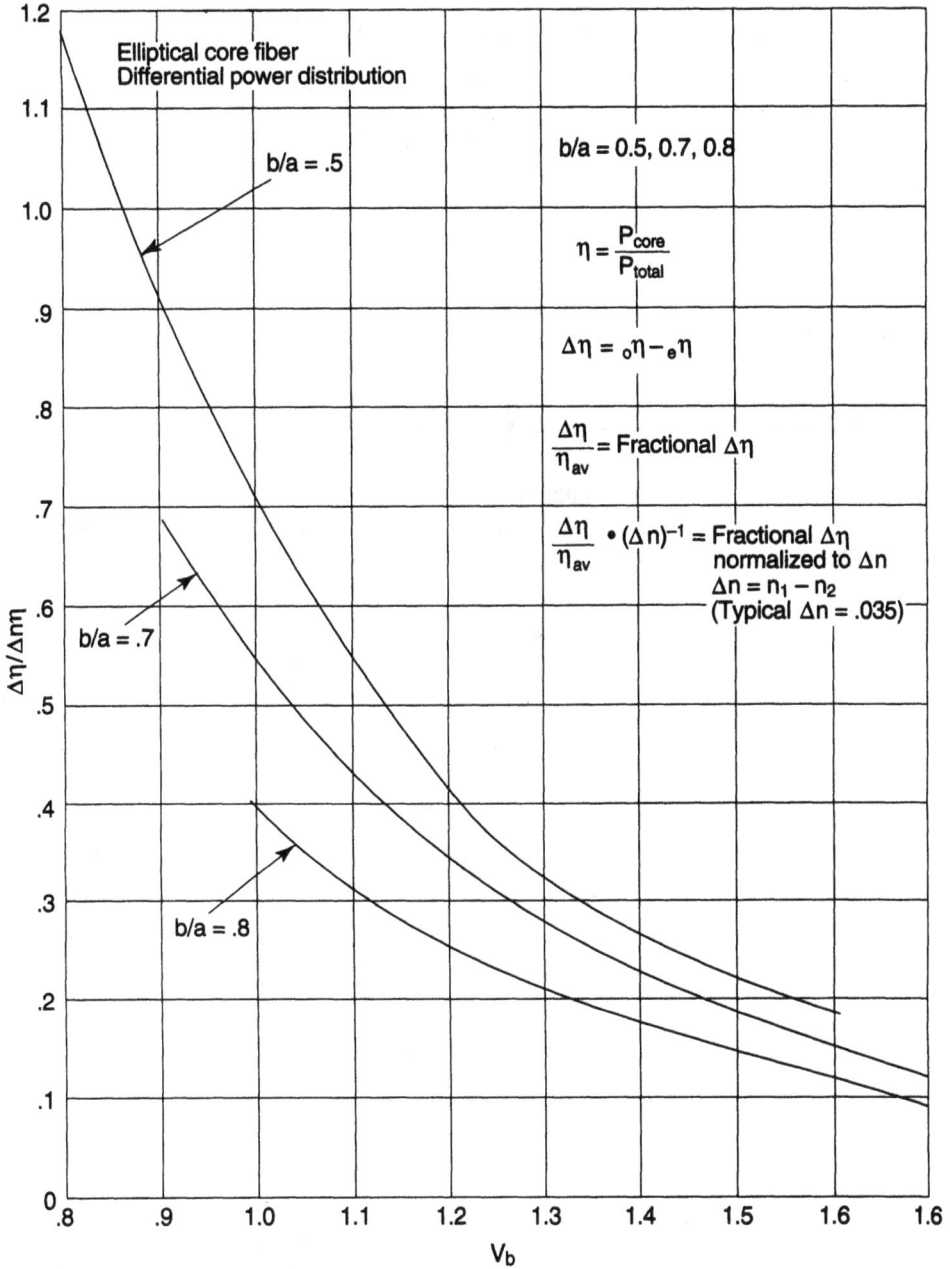

The graph is labeled with the following text and equations:

Elliptical core fiber
Differential power distribution

b/a = .5

b/a = 0.5, 0.7, 0.8

$$\eta = \frac{P_{core}}{P_{total}}$$

$$\Delta\eta = {_o}\eta - {_e}\eta$$

$$\frac{\Delta\eta}{\eta_{av}} = \text{Fractional } \Delta\eta$$

$$\frac{\Delta\eta}{\eta_{av}} \cdot (\Delta n)^{-1} = \text{Fractional } \Delta\eta \text{ normalized to } \Delta n$$
$$\Delta n = n_1 - n_2$$
$$(\text{Typical } \Delta n = .035)$$

b/a = .7

b/a = .8

y-axis: $\Delta\eta/\Delta n\eta$

x-axis: V_b from .8 to 1.6

Figure 9.10 Variation with V_b of the differential power confinement between modes normalized with respect to index difference.

From Figure 9.10, $\dfrac{\Delta\eta}{\eta\Delta n}$ increases dramatically as V_b and therefore η decreases. In order to maximize the effect, it is, therefore, necessary to increase both the ellipticity and the core-cladding index difference and to have most of the power traveling in the cladding. For a practical case of a fiber $b/a = 0.5$, $\Delta n = 0.035$ operating at $V_b = 0.9$

$$\frac{\Delta\eta}{\eta} = 0.031$$

or just over 3%.

9.6 DIFFERENTIAL REFLECTION COEFFICIENT

Another possible cause of differential modal gain is the variation in reflection coefficient at the ends of the fiber due to the different values of effective index $\bar{\beta}$ of the two modes.

The power reflection coefficient R at the interface of the two materials, indices n_1 and n_2 is

$$R = \left[\frac{1 - \dfrac{n_2}{n_1}}{1 + \dfrac{n_2}{n_1}}\right]^2 \tag{9.5}$$

Suppose the surrounding material is in free space, $n_2 = 1$ and n_1 is the equivalent of the fiber waveguide, $\bar{\beta}$. Then

$$R = \left[\frac{1 - \dfrac{1}{\bar{\beta}}}{1 + \dfrac{1}{\bar{\beta}}}\right]^2 \tag{9.6}$$

$$\frac{\partial R}{\partial \bar{\beta}} = 2\left[\frac{1 - \dfrac{1}{\bar{\beta}}}{1 + \dfrac{1}{\bar{\beta}}}\right]\left[\frac{\left(1 + \dfrac{1}{\bar{\beta}}\right)\dfrac{1}{\bar{\beta}^2} + \left(1 - \dfrac{1}{\bar{\beta}}\right)\dfrac{1}{\bar{\beta}^2}}{\left(1 + \dfrac{1}{\bar{\beta}}\right)^2}\right] \tag{9.7}$$

$$\frac{\partial R}{\partial \overline{\beta}} = 2\sqrt{R}\left[\frac{2}{(1+\overline{\beta})^2}\right] \tag{9.8}$$

$$\frac{\partial R}{\partial \overline{\beta}} = \frac{4\sqrt{R}}{(1+\overline{\beta})^2} \tag{9.9}$$

$$\Delta R = \frac{4\sqrt{R}}{(1+\overline{\beta})^2}\Delta\overline{\beta} \tag{9.10}$$

$$\frac{\Delta R}{R} = \frac{4}{\sqrt{R}\,(1+\overline{\beta})^2}\Delta\overline{\beta} \tag{9.11}$$

Figure 9.11 shows $\dfrac{\Delta R}{R}$ versus V_b with $n_1 = 1.485$, $n_1 = 1.450$, and $\Delta n = 0.035$ for ellipticities $b/a = 0.5$, 0.7, and 0.8. Both $\overline{\beta}$ and R move relatively slowly with V_b so that $\dfrac{\Delta R}{R}$ varies mainly with $\Delta\beta$ as can be seen by comparing Figure 9.11 and Figure 4.8. The maximum value of $\dfrac{\Delta R}{R}$ (at approximately $\Delta\overline{\beta}$ max) is still an order of magnitude below the normalized differential confinement (Figure 9.10) so that this latter effect would be the dominant cause of gain anisotropy in a fiber laser with its ends terminated in free space.

9.7 APPLICATIONS

An interesting application of the dual-mode nature of the elliptical core fiber laser is described by Kim et al. [9]. The fiber, pumped with light polarized at 45° to the elliptical axes, lases at two slightly different optical frequencies because of the different propagation constants $_e\beta$ and $_o\beta$ of the two fundamental modes. The optical frequencies mix at the detector to produce a difference frequency f_D, which is sensitive to variations in the difference between the propagation constants (birefringence). The arrangement is shown in Figure 9.12. The birefringence is changed, for instance, by the effects of temperature or strain. The difference frequency f_D is

$$f_D = \frac{\Delta\overline{\beta}}{\overline{\beta}}f_s \tag{9.12}$$

where f_s is the frequency spread of the laser and is related to the linewidth $\Delta\lambda$ by

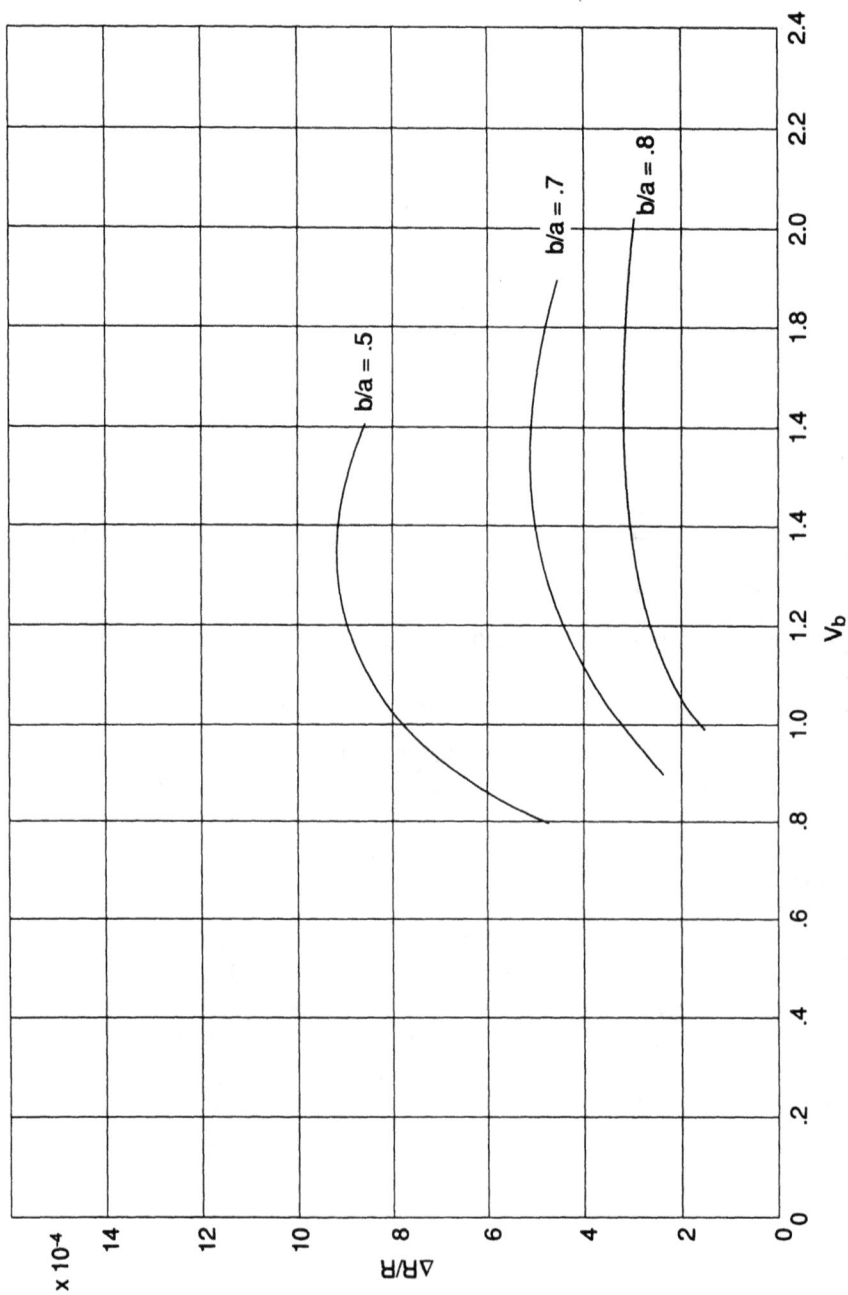

Figure 9.11 Variation with V_b of the normalized differential reflection coefficient between modes.

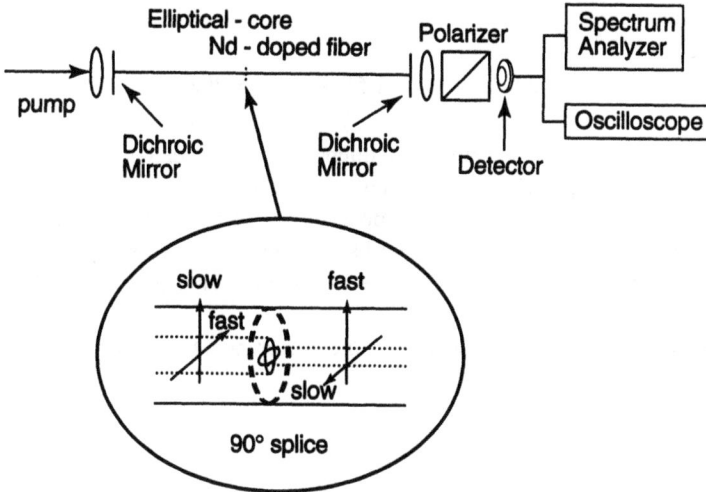

Figure 9.12 Dual mode fiber laser pumped to give two beating optical frequencies.

$$f_s = \frac{c\Delta\lambda}{\lambda^2} \tag{9.13}$$

For example, an Nd-doped fiber with an ellipticity $b/a = 0.5$ and an index difference $\Delta n = 0.035$ is pumped in the fundamental mode region just below cutoff ($V_c = 1.55$) at $V_p = 1.40$ so that at $\lambda_L = 1,090$ nm (the lasing wavelength). The lasing V_L is

$$V_L = V_p \frac{\lambda_p}{\lambda_L} \tag{9.14}$$

the computed average of $\overline{\beta} = \dfrac{{}_e\overline{\beta} + {}_o\overline{\beta}}{2}$ is 1.453 and $\Delta\overline{\beta} = {}_e\overline{\beta} - {}_o\overline{\beta} = 2.3 \times 10^{-4}$ so that

$$f_D = f_s \times \frac{2.3 \times 10^{-4}}{1.453} = 1.58 \times 10^{-4} f_s$$

With the fiber laser linewidth $\Delta\lambda = 10$ nm [9], the bandwidth of the source is

$$f_s = \frac{3 \times 10^8 \times 10^{-8}}{(1.090)^2 \times 10^{-12}} = 2.52 \times 10^{12} \text{ Hz}$$

and f_D becomes

$$f_D = 2.52 \times 10^{12} \times 1.58 \times 10^{-4} = 398 \text{ MHz}$$

In order to reduce the difference frequency f_D, the fiber laser is divided into two unequal lengths. The separate parts are then rejoined with the axes of the elliptical cores aligned major to minor so that the $_e\text{HE}_{11}$ mode of one section couples to the $_o\text{HE}_{11}$ mode of the other and vice versa. If both lengths of fiber were to be equal, the difference frequency would be zero as the sum of the propagation constants for each polarization would be the same.

For unequal lengths, difference ΔL, the difference frequency is proportional to $\dfrac{\Delta L}{L}$ and

$$f_D = \frac{c\Delta\lambda}{\lambda^2} \frac{\Delta\bar{\beta}}{\bar{\beta}} \frac{\Delta L}{L} \qquad (9.15)$$

For a single length of fiber without a splice, the lasing action can be restricted to the $_o\text{HE}_{11}$ mode by bending the fiber around a small radius of curvature [9]. The $_o\text{HE}_{11}$ mode is more closely confined than is the $_e\text{HE}_{11}$ mode so that the bend causes more radiation for the $_e\text{HE}_{11}$ than for the $_o\text{HE}_{11}$, which becomes the dominant lasing mode.

The dual-mode fiber laser, used as a polarimetric sensor thus converts measurement of phase to measurement of frequency, a more accurate and more convenient system.

An optical-fiber switch working on the principle of the change in index induced by the pump in a heavily doped Nd fiber has been described by Sadowski et al. [10]. A probe signal at 633 nm is launched equally into the fundamental and first higher order modes in an overmoded elliptical fiber. The pulsed pump at 807 nm is launched into one of the fundamental modes. The pump, which is tuned to one of the many ground-state absorption resonances, induces a depletion layer of the ground state and population of the metastable level of the Nd ion. This changes the index of the fiber core, which in turn causes the probe modal interference pattern at the output of the fiber to change. The fiber is initially stretched to extinguish one of the lobes of the interference pattern. Pulsing the pump then shifts the pattern from one lobe to another when the induced phase shift between modes is a multiple of π radians.

9.8 SUMMARY

The combination of the doped elliptically cored fiber, the D fiber geometry with its accessible guiding region, and the D loop mirror should have some interesting applications.

REFERENCES

[1] Koester, C. J., and E. Snitzer. "Amplification in a fiber laser," *Applied Optics,* Vol. 3, 1964, pp. 1182–1186.

[2] Kao, K. C., and G. A. Hockham. "Dielectric fiber surface waveguide for optical frequencies," *Proc. IEEE,* Vol. 113, 1966, pp. 1151–1158.

[3] Stone, J., and C. A. Burris. "Neodymium-doped fiber lasers: room temperature c.w. operation with an injection laser pump," *Applied Optics,* Vol. 13, 1974, pp. 1256–1258.

[4] Mears, R. J., L. Reekie, S. B. Poole, and D. N. Payne. "Neodymium-doped silica single-mode fiber." *Electronics Letters,* Vol. 21, 1985, pp. 738–740.

[5] "Selected papers on rare-earth-doped fiber laser sources and amplifiers," Michel Digonnet, Editor, *SPIE Milestone series,* Vol. MS 37, SPIE Optical Engineering Press 1992.

[6] Townsend, J. E., S. B. Poole, and D. N. Payne. "Solution-doping technique for fabrication of rare-earth doped optical fibers," *Electronics Letters,* Vol. 23, 1987, pp. 329–331.

[7] Miller, I. D., D. B. Mortimore, P. Urquhart, J. Ainslie, S. P. Craig, C. A. Miller, and D. B. Payne. "A Nd^{3+} doped c.w. fiber laser using all-fiber reflectors," *Applied Optics,* Vol. 26, 1987, pp. 2197–2201.

[8] Srinivasan, B., S. Gupta, C. Raymond, R. K. Jain, and R. B. Dyott. "Polarization properties of fiber lasers based on rare-earth-doped polarization preserving fibers," *Proc. Conf. Lasers and Electro-optics,* 1994.

[9] Kim, H. K., S. K. Kim, and B. Y. Kim. "Polarization control of polarimetric fiber-laser sensors," *Optics Letters,* Vol. 18, 1993, pp. 1465–1467.

[10] Sadowski, R. W., M.J.F. Digonnet, R. H. Pantell, and H. J. Shaw. "Microsecond optical-optical switching in a neodymium-doped two-mode fiber," *Optics Letters,* Vol. 18, 1993, pp. 927–929.

Appendix A

A.1 MAXWELL'S EQUATIONS

There are four main Maxwell's Equations:

1. The magnetomotive force around a closed path is equal to the conduction current plus the time derivative of the electric displacement through any surface bounded by the path:

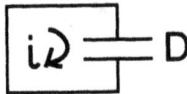

$$\text{Curl } H = i + \frac{\partial D}{\partial t}$$

2. The electromotive force around a closed path is equal to the time derivative of the magnetic displacement through any surface bounded by the path:

$$\text{Curl } E = -\frac{\partial B}{\partial t}$$

3. The total electric displacement through the surface enclosing a volume is equal to the total charge within the volume:

$$\text{Div } D = \rho$$

4. The net magnetic flux emerging from any closed surface is zero.

$$\text{Div } B = 0$$

E = Electric field in strength in volts/meter

H = Magnetic field strength in amps/meter

B = Magnetic flux in webers/meter

D = Electric displacement in coulombs/meter

ρ = Charge density in coulombs/meter^3

i = Current density in amps/meter^2

A.2 PHYSICAL INTERPRETATION OF MAXWELL'S EQUATIONS

Maxwell's Equations are depicted below.

1. An electric current produces a swirl of magnetic field. Current i produces magnetic flux B.

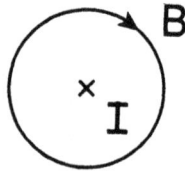

If the current is not concentrated along a single line but is spread out over an area, then the current density in amps/meter^2 produces a magnetic flux B.

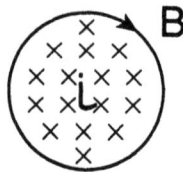

The current need not flow in a conductor but can flow across the dielectric between the plates of a capacitor; that is, if the current is changing with time.

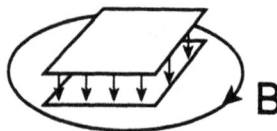

$$i = \frac{\partial D}{\partial t}$$

2. Similarly, a change in the magnetic field with time produces a swirl of electric field (electromagnetic induction).
3. The electric flux emerging from an enclosing surface is due to the electric charge in the surface.

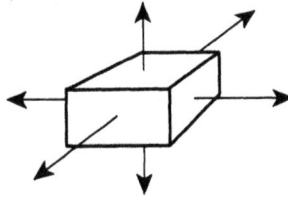

4. The same thing would be true with magnetic flux, but since it is impossible to isolate a magnetic source of flux (i.e., a magnetic pole), the magnetic flux emerging from an enclosed surface is zero.

A.2.1 Equation 1: Magnetic Intensity and Ampères Law

Consider a conductor carrying a current I amps.

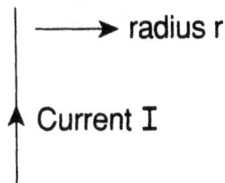

radius r

Current I

Experimentally, the magnetic flux B is proportional to $\frac{\mu I}{r}$, where

$$\mu = \mu' \mu_0$$
$$\mu' = \text{relative permeability}$$
$$\mu_0 = \text{permeability of free space}$$

If $\mu' = 1$ (i.e., in free space), $B = \dfrac{\mu_0 I}{2\pi r}$, $H = \dfrac{I}{2\pi r}$ amps/m, and the integral of H around a closed path $S = \oint \dfrac{I}{2\pi r}\,\delta s$. The integral of δs around a circular path is $2\pi r$ so that $\oint H \cdot \delta s = I$ or the MMF around a closed path is equal to the current enclosed by the path.

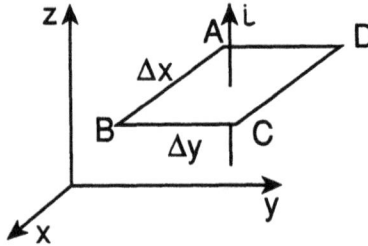

i = current density amps/m^2

Consider a system in rectangular coordinates. A current density i produces a magnetic field strength H.

Let the average value of H_x between A and $B = \widetilde{H}_x$.
Let the average value of H_y between A and $D = \widetilde{H}_y$.
Then, MMF

$$A \rightarrow B = \widetilde{H}_x \Delta x$$

$$B \rightarrow C = \left(\widetilde{H}_y + \frac{\partial H_y}{\partial x}\Delta x\right)\Delta y$$

$$C \rightarrow D = -\left(\widetilde{H}_x + \frac{\partial H_x}{\partial y}\Delta y\right)\Delta x$$

$$D \rightarrow A = -\widetilde{H}_y \Delta y$$

Adding MMFs along all the paths,

$$\text{total MMF around closed path} = \left(\frac{\partial \widetilde{H}_y}{\partial x} - \frac{\partial \widetilde{H}_x}{\partial y}\right)\Delta_x\Delta_y$$

The current flowing through the rectangle $\Delta I = i_z\Delta_x\Delta_y$ so that

$$\left(\frac{\partial \widetilde{H}_y}{\partial x} - \frac{\partial \widetilde{H}_x}{\partial y}\right) \Delta_x \Delta_y = i_z \Delta_x \Delta_y$$

As Δ_x and $\Delta_y \to 0$, $\widetilde{H}_x \to H_x$ and $\widetilde{H}_y \to H_y$
(i.e., the average values are replaced by the values at the point), so:

$$\frac{\partial H_y}{\partial x} - \frac{\partial H_x}{\partial y} = i_z$$

$$\frac{\partial H_z}{\partial y} - \frac{\partial H_y}{\partial z} = i_x$$

$$\frac{\partial H_x}{\partial z} - \frac{\partial H_z}{\partial x} = i_y$$

So far we have only considered conduction currents in a conductor. What happens if there is a break in the conductor?

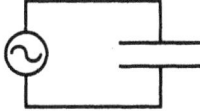

For steady state, Ampères Law still applies as derived. When the current is varying with time, a current flows through the capacitor. Suppose the capacitor has plates of area A separated by distance d. Capacitance $C = \dfrac{\varepsilon_0 A}{d}$, impedance $= \dfrac{1}{\omega C}$ where ω = radian frequency at which the current is varying. If the voltage across the plates is V, then the current I flowing across the capacitor is

$$I = \omega CV = \frac{\omega \varepsilon_0 A V}{d}$$

But $\dfrac{V}{d}$ = electric field E.

Current density $i = \dfrac{\text{current}}{\text{area}} = \omega \varepsilon_0 E = \omega D$

Any change in D with respect to time can be expressed as the sum of a series of sine waves of differing frequencies. Suppose: $D = D_1 \sin \omega t$. Then

$$\frac{\partial D}{\partial t} = \omega D$$

$$i = \frac{\partial D}{\partial t}$$

Ampères Law for currents in conductors and dielectrics becomes

$$\text{Curl } H = i + \frac{\partial D}{\partial t} = i + \varepsilon_0 \frac{\partial E}{\partial t}$$

for free space.

A.2.2 Equation 2: Faraday's Law of Induction

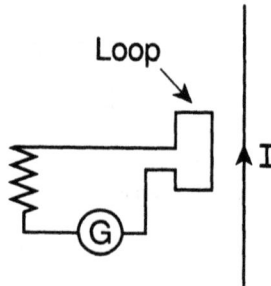

A conductor carries a current I. A magnetic flux links the loop to produce an electromotive force and a current through the ballistic galvanometer G, which has an internal resistance R. The galvanometer deflection depends on the total charge Q put through it:

$$Q = \int_0^t I \, dt = \frac{1}{R} \int_0^t V \, dt$$

The magnetic flux Φ is defined as the time integral of the voltage induced in the loop while the field is being established:

$$\Phi = \pm \int_0^t V \, dt$$

or

$$V = \frac{d\Phi}{dt}$$

V = integral of $E \, ds$ around a closed path = $\oint E \cdot ds$
So that:

$$\oint E \cdot ds = -\frac{d\Phi}{dt}$$

Taking the right-hand side, the flux Φ through a small area da is equal to $B \cdot da$ and

$$-\frac{d\Phi}{dt} = -\frac{\partial B}{\partial t} \cdot da$$

Taking the left-hand side, as the area is made smaller, the integral of $E \cdot ds$ over a closed path divided by the area enclosed by the path is

$$\frac{\oint E \cdot ds}{da} \text{ and becomes Curl E}$$

so that

$$\text{Curl E} = -\frac{\partial B}{\partial t}$$

A.2.3 Equation 3: Gauss's Law

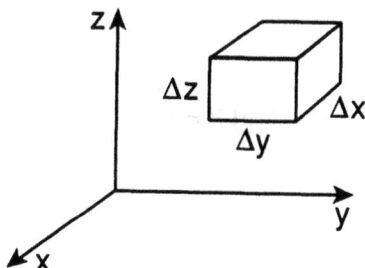

A box of sides Δ_x, Δ_y and Δ_z encloses a small volume. Suppose the electric displacement D varies throughout the volume. Then the total variation is

$$\frac{\partial D_x}{\partial x} + \frac{\partial D_y}{\partial y} + \frac{\partial D_z}{\partial z}$$

Unless there is a source charge within the box, the total variation in D must be zero, or

$$\text{divergence of } D = 0$$

If there is a source of charge density ρ coulombs/meter^3 in the box, then

$$\frac{\partial D_x}{\partial x} + \frac{\partial D_y}{\partial y} + \frac{\partial D_z}{\partial z} = \rho$$

In free space there are no sources of charge, so

$$\text{Div } D = 0$$

A.2.4 Equation 4

The net magnetic flux emerging from a closed surface is zero:

$$\frac{\partial B_x}{\partial x} + \frac{\partial B_y}{\partial y} + \frac{\partial B_z}{\partial z} = 0$$

This is always true since it is impossible to isolate a source of magnetic flux.

A.2.5 Summary

$$\text{Curl } H = \frac{\partial D}{\partial t} + i$$

$$\text{Curl } E = -\frac{\partial B}{\partial t}$$

$$\text{Div } D = \rho$$

$$\text{Div } B = 0$$

$$\nabla \times = \text{Curl} = \frac{\text{line integral around a small area}}{\text{small area}}$$

$\nabla\cdot = \text{Div} = $ flux per unit volume flowing out of a unit volume

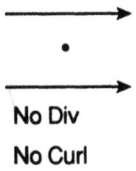

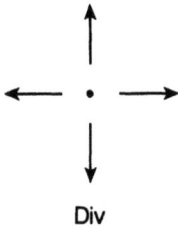

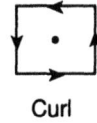

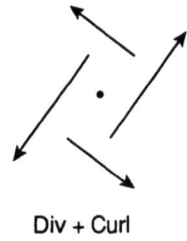

No Div
No Curl

Div

Curl

Div + Curl

Appendix B

B.1 ELECTROMAGNETIC WAVES IN FREE SPACE

There are three supplementary equations to the four main Maxwell's Equations:

$$B = \mu H$$

$$D = \varepsilon E$$

$$i = \sigma E \text{ (Ohm's Law, } \sigma = \text{conductance)}$$

In free space,

$$\mu_0 = 1.257 \times 10^{-6} \text{ henry/meter} = 4\pi \times 10^{-7}$$

$$\varepsilon_0 = 8.55 \times 10^{-12} \text{ farad/meter} = \frac{1}{36\pi \times 10^9}$$

Substituting these equations in the four main ones,

$$\nabla \times E = -\mu \frac{\partial H}{\partial t}$$

$$\nabla \cdot H = 0$$

$$\nabla \times H = \sigma E + \varepsilon \frac{\partial D}{\partial t}$$

$$\nabla \cdot D = \rho$$

B.2 HARMONIC FIELDS

If the fields are varying sinusoidally (and any variation can be analyzed as a series of sinusoidal variations of different frequencies); that is,

$$E = E_1 e^{j\omega t}$$

$$H = H_1 e^{j\omega t}$$

then $\dfrac{\partial H}{\partial t} = j\omega H$ and $\dfrac{\partial E}{\partial t} = j\omega E$, so that

$$\nabla \times E = -j\omega\mu H$$

$$\nabla \cdot H = 0$$

$$\nabla \times H = \sigma E + j\omega\varepsilon E = (\sigma + j\omega\varepsilon)E$$

$$\nabla \cdot D = \rho \text{ or } \nabla \cdot E = \frac{\rho}{\varepsilon}$$

The loss represented by σ is often included in a complex $\varepsilon = \varepsilon'$. If $\varepsilon' = \dfrac{\sigma}{j\omega} + \varepsilon$ then

$$\nabla \times H = j\omega\varepsilon' E$$

Eliminating E or H from the equations gives

$$\frac{\partial^2 E_x}{\partial x^2} + \omega^2 \mu\varepsilon' E_x = 0, \quad \frac{\partial^2 H_x}{\partial x^2} + \omega^2 \mu\varepsilon' H_x = 0$$

$$\frac{\partial^2 E_y}{\partial y^2} + \omega^2 \mu\varepsilon' E_y = 0, \quad \frac{\partial^2 H_y}{\partial y^2} + \omega^2 \mu\varepsilon' H_y = 0$$

$$\frac{\partial^2 E_z}{\partial z^2} + \omega^2 \mu\varepsilon' E_z = 0, \quad \frac{\partial^2 H_z}{\partial z^2} + \omega^2 \mu\varepsilon' H_z = 0$$

These are equations of a wave in a medium whose ε and μ do not depend on H, E, or ω and which obey Ohm's Law. A more complicated solution exists, for instance, for the propagation of EM waves through an ionized gas, which does not obey Ohm's Law.

B.3 CLASSIFICATION OF SOLUTIONS TO THE WAVE EQUATIONS

The solutions are classified as follows:

1. No longitudinal wave (that is a wave with E and H both in the same direction) exists in free space. Suppose such a wave did exist propagating in z direction.

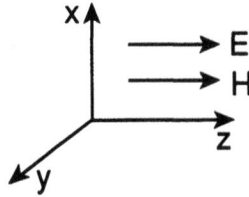

All the other components of E and H fields are zero, $E_x = 0$, $E_y = 0$, $H_x = 0$ and $H_y = 0$. Then

$$(\nabla \times E)_z = \left(\frac{\partial E_y}{\partial x} - \frac{\partial E_x}{\partial y}\right) = 0$$

But $\nabla \times E_z = -j\omega\mu H_z$ and therefore $H_z = 0$ and

$$(\nabla \times H)_z = \left(\frac{\partial H_y}{\partial x} - \frac{\partial H_x}{\partial y}\right) = 0$$

But $\nabla \times H_z = -j\omega\varepsilon' E_z$ and therefore $E_z = 0$.

2. Waves with electric and magnetic fields both in the transverse to the direction to propagation and with $E_z = 0$; $H_z = 0$ are transverse electromagnetic (TEM) waves.
3. Waves with only a component of electric field in the direction of propagation are electric E waves or transverse magnetic (TM) waves.
4. Waves with only a component of magnetic field in the direction of propagation are magnetic H waves or transverse electric (TE) waves.

B.4 UNIFORM TEM PLANE WAVE

Plane: Surfaces of same phase are planes.
    Uniform: Amplitude is the same along equiphase planes.
    Suppose the XY plane is the equiphase plane

$$\frac{\partial E}{\partial x} = 0, \quad \frac{\partial H}{\partial x} = 0, \quad \frac{\partial E}{\partial y} = 0, \quad \frac{\partial H}{\partial y} = 0$$

that is, there is no variation of E or H over the XY plane. Also $E_z = 0$ and $H_z = 0$ with no E or H in the direction of propagation. Then

$$(\nabla \times E)_y = \frac{\partial E_x}{\partial z} - \frac{\partial E_z}{\partial x} = -j\omega\mu H_y$$

$$(\nabla \times H)_x = \frac{\partial H_z}{\partial y} - \frac{\partial H_y}{\partial z} = (\sigma + j\omega\varepsilon)E_x$$

Substituting for the quantities that are zero,

$$(\nabla \times E)_y = \frac{\partial E_x}{\partial z} = -j\omega\mu H_y$$

$$(\nabla \times H)_x = \frac{\partial H_y}{\partial z} = (\sigma + j\omega\varepsilon)E_x$$

Differentiating with respect to z,

$$\frac{\partial^2 E_x}{\partial z^2} = -j\omega\mu\frac{\partial H_y}{\partial z}; \qquad \frac{\partial^2 H_y}{\partial z^2} = (\sigma + j\omega\varepsilon)\frac{\partial E_x}{\partial z}$$

$$\frac{\partial^2 E_x}{\partial z^2} = -j\omega\mu(\sigma + j\omega\varepsilon)\,E_x; \qquad \frac{\partial^2 H_y}{\partial z^2} = -j\omega\mu(\sigma + j\omega\varepsilon)\,H_y$$

Solving these equations,

$$E_x = Ee^{-\gamma z}e^{j\omega t}$$

$$H_y = He^{-\gamma z}e^{j\omega t}$$

$$\gamma = \pm\left[j\omega\mu(\sigma + j\omega\varepsilon)\right]^{\frac{1}{2}} = \pm\left[j\omega\mu\sigma - \omega^2\varepsilon\mu\right]^{\frac{1}{2}}$$

γ is known as the propagation constant $\gamma = \alpha + j\beta$.
α = real part and represents attenuation with distance.
$j\beta$ = imaginary part and represents phase constant.

$$E_x = Ee^{-(\alpha+j\beta)z}e^{j\omega t}$$

Put $\beta = \dfrac{2\pi}{\lambda}$, E_x repeats at periodic intervals of λ (i.e., wavelength).

B.5 VELOCITY OF WAVE

$$v = f\lambda = \frac{\omega}{\beta} = \frac{2\pi f}{2\pi/\lambda}$$

$$\text{For no loss, } \sigma = 0; \quad \beta = \omega\,(\varepsilon\mu)^{\frac{1}{2}}; \quad v = \frac{1}{(\varepsilon\mu)^{\frac{1}{2}}}$$

$$\text{Freespace velocity} = \frac{1}{(\varepsilon_0\mu_0)^{\frac{1}{2}}}$$

Relationship Between E_x and H_y

$$H_y = -\frac{\gamma}{j\omega\mu}E_x$$

$$\frac{E_x}{H_y} = -\frac{j\omega\mu}{\gamma} = \left[\frac{j\omega\mu}{(\sigma + j\omega\varepsilon)}\right]^{\frac{1}{2}}$$

If $\sigma = 0$

$$\frac{E_x}{H_y} = \left[\frac{\mu}{\varepsilon}\right]^{\frac{1}{2}} = \text{wave impedance } Z_w$$

For free space,

$$Z_w = \left[\frac{\mu_0}{\varepsilon_0}\right]^{\frac{1}{2}} = [4\pi \times 10^{-7} \times 36 \times 10^9]^{\frac{1}{2}} = 120\pi \simeq 377\Omega$$

Impedance of free space = 377Ω.

Appendix C

Table C.1

Refractive Index of Germania-Doped Silica

Germania Mol. Fr.	$n_{(810)}$	$n_{(1.300)}$	$n_{(1.500)}$
0	1.4531	1.4469	1.4446
0.05	1.4600	1.4537	1.4514
0.10	1.4670	1.4605	1.4583
0.15	1.4739	1.4674	1.4651
0.20	1.4809	1.4743	1.4720
0.25	1.4879	1.4812	1.4790
0.30	1.4950	1.4882	1.4859
0.35	1.5021	1.4952	1.4929
0.40	1.5092	1.5022	1.5000
0.45	1.5164	1.5093	1.5071
0.50	1.5236	1.5164	1.5142
0.60	1.5382	1.5307	1.5285
0.70	1.5529	1.5452	1.5431
0.80	1.5678	1.5599	1.5577
0.90	1.5828	1.5747	1.5726
1.0	1.5981	1.5897	1.5876

Table C.2
Refractive Index of Fluorine-Doped Silica

Fluorine Mol. Fr.	$n_{(810)}$	$-\Delta n$	$n_{(1.300)}$	$-\Delta n$	$n_{(1.500)}$	$-\Delta n$
0	1.4531	0	1.4469	0	1.4446	0
0.002	1.4522	0.0009	1.4460	0.0009	1.4437	0.0009
0.004	1.4513	0.0018	1.4451	0.0018	1.4428	0.0018
0.006	1.4504	0.0027	1.4442	0.0027	1.4419	0.0027
0.008	1.4495	0.0037	1.4433	0.0036	1.4410	0.0036
0.010	1.4486	0.0046	1.4424	0.0045	1.4402	0.0045
0.012	1.4477	0.0055	1.4415	0.0054	1.4393	0.0054
0.014	1.4468	0.0064	1.4406	0.0063	1.4384	0.0062
0.016	1.4458	0.0073	1.4397	0.0072	1.4375	0.0071
0.018	1.4449	0.0082	1.4389	0.0081	1.4366	0.0080
0.020	1.4440	0.0091	1.4380	0.0090	1.4357	0.0089

Index

The Artech House Optoelectronics Library

Brian Culshaw, Alan Rogers, and Henry Taylor, *Series Editors*

Optical Fiber Sensors, Volume II: Systems and Applicatons, John Dakin and Brian Culshaw, editors

Optical Interconnection: Foundations and Applications, Christopher Tocci and H. John Caulfield

Optical Network Theory, Yitzhak Weissman

Optical Transmission for the Subscriber Loop, Norio Kashima

Reliability and Degradation of LEDs and Semiconductor Lasers, Mitsuo Fukuda

Semiconductor Raman Laser, Ken Suto and Jun-ichi Nishizawa

Semiconductors for Solar Cells, Hans Joachim Möller

Single-Mode Optical Fiber Measurements: Characterization and Sensing, Giovanni Cancellieri

For further information on these and other Artech House titles, contact:

Artech House
685 Canton Street
Norwood, MA 02062
617-769-9750
Fax: 617-769-6334
Telex: 951-659
email: artech@world.std.com

Artech House
Portland House, Stag Place
London SW1E 5XA England
+44 (0) 71-973-8077
Fax: +44 (0) 71-630-0166
Telex: 951-659
email: bookco@artech.demon.co.uk